The Obama Legacy

The Obama Legacy

by

Earl Ofari Hutchinson

MIDDLE PASSAGE PRESS

The Obama Legacy

Printed in the United States

Published by
Middle Passage Press
5517 Secrest Drive
Los Angeles, California 90043

Indexed by Barbara Bramwell Hutchinson
Designed by Alan Bell

Publisher's Cataloging-in-Publication
(Provided by Quality Books, Inc.)
Hutchinson, Earl Ofari, author.
The Obama legacy / by Earl Ofari Hutchinson.
pages cm
Includes bibliographical references and index.
LCCN 2016919942
ISBN 978-0692813096
1. Obama, Barack.
2. Obama, Barack—Influence.
3. Presidents—United States—History—21st century.
4. African Americans—Politics and government.
5. United States—Politics and government—2009-
I. Title.
E907.H8825 2017
973.932
QBI16-900086

Table of Contents

The Obama Legacy

Introduction

"Cancel every unconstitutional executive action, memorandum and order issued by President Obama." President elect Donald Trump was blunt when he announced that rescinding Obama's executive orders would be his first order of business in his first 100 days in the White House. This was not the first-time Trump vowed to dump Obama's executive orders in the dustbin. He shouted the same thing to an audience at a campaign rally in Mesa, Arizona in December, 2015. "I walk in, sign, I don't have to go through Congress."

Trump wasn't the only one who took a direct bead on dismantling Obama's administrative and legislative initiatives. GOP House Speaker Paul Ryan also wasted no time in announcing that Congress, now controlled by the GOP, would move quickly to try and invoke the rarely used Congressional Review Act. This gives Congress the power to toss aside anything that Obama might try to push through in the waning days of his tenure in the White House, as well as nearly 200 regulations that he had signed off on in the last year of his presidency. The first targets almost certainly would be the rules from the Environmental Protection Agency, reducing greenhouse gas emissions from commercial trucks and buses, and from the Labor Department raising the salary threshold for workers eligible for overtime pay from $455 a week to $913 a week.

Ryan didn't stop there. He flatly warned Obama: "Then administrations would know that they will be held accountable, even after they leave office." Trump almost certainly would gleefully quickly rubber stamp the Act.

This was not idle and partisan saber rattling of Obama to ensure that he took no action that Trump and the GOP had routinely reviled for virtually every moment of his eight years in the White House. This was the opening gun of what would almost certainly be a full-bore effort to roll back, if not outright dismantle, Obama's legacy.

The irony was that Trump and the GOP weren't the first to take a shot at Obama's legacy. Some Democrats did that four years before Trump. During the run-up to the 2012 presidential contest. Obama was expected to be in a tough race with GOP presidential contender Mitt Romney. His popularity numbers had seesawed up and down in the year before the election, making Democrats nervous. The excuses some of them gave for their chill toward backing Obama's reelection bid would have filled up a legal pad.

The charge was that he made much too nice with the GOP. He allegedly put Medicare and Social Security on the chopping block during budget negotiations in 2010 and 2011. He didn't push aggressively enough for a full-blown FDR-style jobs program. He let Wall Street and banks off the hook with a placid, terribly compromised financial reform bill. He didn't do enough on home foreclosures. The conservative Blue Dog and moderate Democratic congresspersons and senators that represented shaky swing and conservative districts were scared stiff that if they rubbed shoulders too close to Obama they would be signing their political obituary for reelection.

Their frost toward Obama was far more worrisome than the pesky, nuisance criticisms of Ralph Nader about finding some progressive, pro-labor Democrat to run against Obama in 2012. This fed the naive and delusional thinking of some radicals that a challenge to Obama would have somehow shoved him and the Democratic party to embrace an unabashed anti-corporate, anti-war, anti-poverty, pro-union, bank and financial crackdown agenda. This talk quickly faded into the news dust bin. It was revived for a brief-moment when it seemed that Occupy Wall Street, with its nationwide anti-wealth inequality camp-ins in 2011, might become an organized movement with visible leadership, tangible goals. And that it might target Obama as much as protesters targeted the corporations and GOP for aiding and abetting corporate pillage.

* * * * *

This didn't happen. However, given the negative view of Obama's performance, if it had ever amounted to anything beyond grumbles, the effect would have been to create party paralysis and division at worst, and uncertainty at best. This would have been disastrous to a presidential campaign.

That was amply proven when Ronald Reagan challenged President Gerald Ford in 1976 and when Senator Ted Kennedy challenged President Jimmy Carter in 1980. Their challenges weakened both presidents, divided the party, and ultimately helped make possible Carter's win over Ford and Reagan's win over Carter.

The inescapable conclusion that voters would have drawn from this was that Obama's GOP opponent might have something better to offer voters on the crucial make or break issues of jobs and the economy. This was especially dangerous with polls consistently showing that a solid majority, including a lot of Democrats, gave Obama a low mark on his handling of the economy. The other great danger in the Democrat's push back from the President was that it would have watered down even more the critical enthusiasm level for Obama.

This was the biggest factor that powered him to the White House in 2008. Independents and youth voters were fired up by Obama's message of hope and change, and fed up with the GOP's corruption, bungling, blatant cronyism and scandals, and George W. Bush's fumbles and ineptitude. They stampeded to the polls in droves to back Obama. This made the crucial difference in the must-win swing states of Ohio, Florida, Pennsylvania, North Carolina, and Virginia. Bush won three of these five states in 2000 and 2004. Obama won all five in 2008.

This little pre-election drama underscored the point that the burdens on an incumbent president were terrible, and unfair, but real. That was what Obama had to contend with in 2012 and in virtually every other year of his White House tenure. It also underscored just how fragile the legacy of a president could be, since the actions he took in the Oval Office and the approval of, and opposition to, those actions and the public's perception of the worth of those actions were ultimately judged.

Obama's legacy was wildly and admiringly applauded and praised and just as wildly and un-admiringly assailed and vilified. But what is his legacy? Historians have yet to assess that legacy. That will take the passage of years. What we can do, though, is examine the program and policies that Obama embraced and battled for. Those actions affected millions of Americans. Their lasting impact will be felt, debated, and dissected long after he is out of office. The word "legacy" in the dictionary definition is a bequest, something left to those who come after, something to be remembered. *The Obama Legacy* makes a deep, balanced, wide-ranging probe of the big-ticket issues that marked his years in office and presents as well selections from his signature speeches. It places the Obama White House squarely in the context of what he hoped to achieve in the Oval Office and what he, as other American presidents, want be remembered for.

CHAPTER 1

Countering the Bush Failure

President Obama had barely had time to get situated in the Oval Office when many liberal critics screamed that he was backpedaling on his promises to totally dismantle the most odious of his predecessor, George W. Bush's policies. They were ticked at him because he did not dump all of Bush's restrictions on stem cell research, eliminate the discrimination encoded in the faith based programs, downplayed talk of Buy American, and would not haul Bush officials accused of condoning torture and illegal wiretapping into a court docket, and did not drastically toughen regulations on Wall Street banks and brokerage houses.

Certainly, Obama made many promises when he was in a hot race with Hillary Clinton to win the Democratic presidential nomination in 2008. He made more promises when he was in a seesaw battle early on with Republican rival John

McCain to bag the White House. He simply did what all candidates do who want to win. He made promises, lots of promises, that he could not possibly fulfill, at least not right away, and not in the way those who literally took him at his word wanted him to fulfill. The big difference between a presidential candidate and a president was that the candidate did not have to work with Congress, Wall Street, corporations, the Defense establishment, and, most importantly, the other party, to get legislation passed.

Presidential candidates often sound like wild men on the campaign trail and then quickly settle into good centrists once in the White House. They invariably litter the trail with bruised and hurt feelings, dashed hopes, and they hear shouts of betrayal. Obama heard that sooner than most, in part because liberals put a blind faith belief in his promise to make instant change. And in greater part because the horror of Bush's policies was so great, they morphed Obama into the near mythical man on the white horse who could magically right all political wrongs. Obama saw this coming and he did everything he could to damp down expectations that he could make quick changes. But he didn't have to do that.

From the start, Obama was a pragmatic, centrist Democrat. If he had been anything else he could never have gotten the stamp of approval from top Democrats, broken the cash registers on fund raising during his campaign, beat down the Clinton Machine, gotten the parade of endorsements from former Reagan and Bush Sr., and even Bush Jr. officials. He could not have drawn the raves of virtually every major news outlet if there was even the slightest hint that he was a populist thumping crusader.

Obama's centrist bent was plainly evident during the campaign when he and Republican rival John McCain at times sounded like they were in more agreement than not on the issues of expansion of stem-cell research, immigration, faith-based social services, expanded government wiretapping, building more nuclear power plants, global warming, fair trade, and the death penalty. The similarity between the two was more glaring when Obama edged closer to McCain on their plans on health care and taxes and the Iraq War. This did not mean that Obama was a McCain in sheep's clothing. It just meant that he had to find a middle ground on these issues to guarantee that McCain and GOP conservatives did not do everything they could to torpedo his programs when he got in the White House.

* * * * *

The truism in American presidential politics is that liberals and even one time progressives always run to the political left in the early stages of a campaign. They then move quickly to the center or even rightward as victory becomes a real possibility.

Even when Obama spoke most passionately about change he kept the door wide open to reshape, massage, and contour policy issues to conform to what was pragmatic, doable and acceptable. He had the standard liberal reservations that the Wall Street bailout gave too much away to bad behaving, profligate spending CEOs of the big banks and that massive tax cuts wouldn't do much to help the poor and the middle class unemployed.

However, he backed the bailout and tax cuts, albeit modified, to get the support of a handful of congressional Republicans. He opposed the Iraq war, but if he pushed for a quick and immediate withdrawal, military conservatives would have howled and dug in their heels to resist. He pragmatically talked about flexible timetables and a phased withdrawal from Iraq.

Eventually he would implement the Iraq withdrawal, as well as parts of other policy items that he promised to change in an effort to confront Bush's failures. This was the only way he could assure liberal critics and the millions that backed him that he was different from his predecessor. This would, he hoped, send the strong signal that he'd deliver on the hope and change that he promised his administration would bring.

* * * * *

However, that didn't answer this burning question. In the overall political scheme of things did the Bush administration do more or less for the country in every vital area from defense to the economy than the Obama administration? Historians and scholars will endlessly debate that question as they have done in making other comparisons of the effectiveness or incompetence of a president both within and without office with other presidents. The Bush versus Obama presidencies are no different. The start and end-point is simply to look at the check list of issues and needs each administration faced, crunch the numbers, and try to make sense of those numbers in terms of its overall effect on the nation. In doing that there are some surprises and myth shattering in comparing Bush and Obama.

Contrary to belief and perception, Obama actually did spend more on defense than Bush, to the tune of more than $200 billion. This didn't mean that Obama was spending more to wage war. Quite the contrary, the spending went up even while he was drastically scaling back U.S. military involvement in Iraq and Afghanistan. The spending hike was due to the continued bloat in the cost of military hardware, and training, the global maintenance of US bases, and personnel, health, pension and insurance costs, as well as increased spending on intelligence gathering and technology improvements for the military.

Obama and Bush both faced a near crisis free fall in the economy. They approached the crisis in far different ways. Bush relied almost exclusively on tax cuts to try and prime the economic pump. Those cuts and rebates totaled billions and were roundly ripped as a giveaway to the wealthy and big corporations and for ballooning the deficit. Obama pushed hard to spend more on job creation and training and tried with every budget he proposed to up the spending in this area. In almost all cases he met hard resistance from the GOP controlled House which demanded bigger and bigger budget cuts allegedly to stave off inflation and bring down the deficit that Bush's tax cuts had blown up. The contrast between how both approached the tank in the economy was stark when momentarily Democrats controlled both houses of Congress in 2009. Obama then managed to get through a near $800 billion economic stimulus plan that directly created jobs in education and infrastructure. He also used bailout funds to help subsidize homeowners who were under water with their home mortgages and faced the prospect of being plopped on the street with foreclosure.

The single biggest drain on the economy and beyond was exploding health care costs and the inability of millions to get health care at any cost, let alone affordable health care. Bush took a stab at damping down costs when his administration established the Medicare Part D prescription drug program. This cut some drug costs for seniors. But it also further ballooned the deficit. By the time Obama took office in 2009, the number one cause of bankruptcies in the country was health care costs, mostly due to the astronomical costs of treating and paying for catastrophic illnesses. His Affordable Care Act, Obamacare, aimed to cut the costs by placing the emphasis on prevention before illnesses became catastrophic. It would take time to assess just how effective the Act was in damping down medical costs. However, for the millions that obtained health care access for the first time, it was a boon.

Obama, as did Bush, put a lot of emphasis on free trade agreements. Both did a lot of globetrotting to help seal deals with many countries and regional trade associations. The big difference was that Obama's goal was always how these agreements could create more jobs. He took much heat from labor and progressive groups for pushing the Trans Pacific Partnership in that it gave away too much in lax environmental and labor protections, was loaded with goodies for corporations and would do little to bring jobs back to this country. The pros and cons of this was a moot point since Congress wasn't likely to approve the pact anyway, at least not before Obama left office in January 2017.

* * * * *

The GOP's long standing knock against Democrats was that they are the worst thing in the world for business by hamstringing business and investors with an endless labyrinth of governmental red tape and regulations. It hit Obama hard with the standard charge. It got even worse when Obama firmly backed tighter regulations to rein in Wall Street and the Big Banks for their casino style lending and speculating practices which drove the financial industry to the brink of ruin in 2008. It sent millions of Americans into a downward spiral when plants and businesses went under, and joblessness soared. However, by continuing to push for tighter financial controls through the Dodd-Frank Wall Street Reform Act and the Consumer Financial Protection Agency, Obama at least insured that Wall Street could not run completely rough shod over consumers.

Obama did spend a lot while in office. He was hammered repeatedly for running the deficit to record numbers over Bush's deficit spending numbers. The big increase had much to do with trying to revive the economy and end the

Great Recession; a recession that Obama didn't create, but inherited from Bush. There is also the hard reality that deficits are created not solely by president who advocate profligate spending, but by factors that presidents have little control over; namely sustained business and economic growth that comes in cycles.

When the cycle turns upward, revenues rise and deficits fall. During Obama's second term the economy recovered, and the deficit steadily dropped. However, the big knock against Obama was that he. like Bush, continued to push the debt ceiling higher and higher. He did, but the nation's debt is a cumulative figure, in part due to spending costs inherited from prior administrations, and in greater part because of the plunge in federal revenues during the Great Recession from 2008-2010, the outlandish Bush tax cuts that sharply reduced tax receipts, and the seemingly never-ending increase costs of Medicare, Social Security, the legions of mandatory government spending programs, and of, course, the continued costs of the Bush initiated Iraq and Afghanistan wars, and war on terror.

The GOP and the pack of chronic Obama bashers loved to revile him for a big and costly government, while trying to make the case that things were much better with spending, the deficit, debt and the economy under Bush. This was pure political saber rattling and one-upmanship aimed at Obama that deliberately muddled facts and history and the Obama record versus Bush. It couldn't muddle this, though. Obama had the unenviable job of trying to clean up the messes that

Bush made of everything from the economy to foreign policy. At the end of his two terms the success he had in doing that stood in stark contrast to the failures of his predecessor who made the messes.

CHAPTER 2

The Obama Care Presidency

Two weeks after he was sworn in to office in January, 2009, Obama was frank. He said that if he didn't deliver he'd be "a one term proposition." He knew that as the nation's first African-American president, millions were watching to see what he could and would deliver. Millions of others eagerly hoped against hope he'd fail miserably. He was under the white-hot glare of the public to deliver the goods, or be quickly dumped in the presidential has-been.

New presidents always ride into office on the crest of both voter hopes and euphoria about the prospect of change and disgust at and voter fatigue with the former seat warmer in the White House. And new presidents just as quickly see their approval ratings dip or freefall. It was easy to see why. They try to do too much too soon; promise not to do political business in the old ways; try to make too drastic legislative changes, or quickly reverse the bad old policies of their predecessor. It was the fabled man on the white horse coming to the rescue. This was, of course, just that, fable. Real politics and an impatient public knock that storybook notion for a loop.

In Obama's case, he gambled that his presidency would be a crowning success if he could beat back the fine-tuned, well-oiled and well-endowed health care industry juggernaut and get health care reform. That meant getting real health care reform through Congress and into law. Only one president had been able to do that. That was Lyndon Johnson. He arm-twisted, browbeat, and out-smarted Congress and the health care industry to get Medicare. Johnson had won a landslide election victory in 1964, had monumental, hard-nosed political skills and had the reform spirit of the civil rights movement and a solid Democratic party behind him. He had the wellspring of public sympathy after JFK's murder. Obama was not LBJ, politically. He had neither the times or Johnson's massive mandate for change going for him.

Above everything else, the voters put Obama in the White House to make the economy right, rein in the Wall Street greed merchants, save jobs and homes, and get the credit pipeline to businesses open. That didn't happen to the extent it could and should have. Instead they got a raucous and contentious health care reform fight that gave a badly fractured and reeling GOP, which had been the butt of scorn and jokes after Obama's 2008 victory, something that it never dreamed in its wildest dreams could happen. That was the weapon to get back in the political hunt.

Despite that, Obama eventually got a health care bill to sign. The Affordable Care Act was signed into law on March 23, 2010. The law, though, remained shrouded in controversy. Progressives screamed that the bill was stripped of a public option, and was a deal laden with giveaways and concessions to the big pharmaceuticals. The GOP screamed louder that the law taxed and compelled far too many to get a costly, and maybe even unwanted, health care package.

This was one of the relentless attack points that the GOP dredged up in endless committee hearings, a Supreme Court challenge and dozens of votes to repeal the Act. They all failed. Obamacare stood the challenge for the very reasons that it became law. It provided tens of millions of Americans with access to a health plan, there would be subsidies for low-income persons to offset the costs, a half million children would be covered under their parent's plans, and millions of dollars would be allocated for research and testing. It also established more than 1000 new health care facilities in many rural and urban communities, and the National Health Service Corps workforce was tripled. Millions of elderly and disabled Americans covered under Medicare were now provided with no cost access to health care preventive services.

Most important was that the millions of chronically uninsured and underinsured Americans for the first time had full access to health care. The Congressional Budget Office estimated that the number of uninsured Americans would fall from fifty-five million to thirty-one million between 2010 and 2023. The proportion of the population with coverage would rise from eighty-two per cent to ninety-two per cent. This meant little to the GOP critics.

* * * * *

The GOP hit plan on Obamacare was always much more than a battle to scrap a health care plan that it reviled as flawed, failed, and a big government intrusion into the lives of Americans. It was a crude and cynical political ploy to inflame millions of Americans and stoke hostility to Obama and the Democrats. Polls, however, at the time of the fierce battles over Obamacare and afterwards consistently showed that while many Americans had deep concerns and even legitimate fears about the short-term effect of Obamacare, they never wanted it scrapped. The GOP, and later, GOP presidential contender Trump's drumbeat hit campaign on Obamacare didn't erase that fact. Presidents Franklin D. Roosevelt, Harry Truman, LBJ, and Bill Clinton, tried, to varying degrees, to get a national health care bill into a law. They failed. Obama didn't.

However, the danger signs were there. Trump repeatedly parroted the standard GOP attack line during the 2016 presidential campaign; that Obamacare should be repealed and replaced. The first pronouncement he made after his election was that he'd move quickly to take action on it. However, it wasn't that simple. And Trump knew it. He immediately reversed gears and hedged that there were some parts of Obamacare that may need to stay in place. This was back-door recognition of a hard-political fact.

By the time Trump took office, more than 20 million persons had been added to the health care roles under Obamacare, and just lopping them off would court political disaster. The brutal political reality was that Obamacare had made a difference in the lives of many Americans. Despite Trump and the GOP's bluster about junking it, the chances were good that the Affordable Care Act would stand up under the GOP assault. This was a lasting testament to Obama.

CHAPTER 3

Making Like FDR

The comparison of then Democratic presidential candidate Obama in 2008 to Franklin Delano Roosevelt at the height of the presidential campaign was hyped, overblown and made mostly to sell magazines, puff up TV pundit sound bites, and by a few over enthusiastic Democratic party campaign boosters. Though undoubtedly flattered by it, candidate Obama did not encourage the comparison to FDR. Many hoped that he would inch toward looking and acting like FDR. They relentlessly pushed, prodded, and hectored him to lurch in that direction. There were many days to come of bitter frustration and disappointment, punctuated by loud grumblings of betrayal. But how realistic was the Obama-FDR nexus expectation? Obama, as FDR, knew that he was in a political life and death, take-no-prisoners war with his political enemies— the

GOP, ultra conservative Democrats, Wall Street, and the big bankers and big manufacturers. Unlike FDR, for months, he soft peddled, coddled, and placated his opponents even as they made absolutely no effort to mask their loathing of his policies and presidency, and made it abundantly clear they would stop at nothing to hound him from office. FDR, by contrast, hit back hard at his enemies as obstructionists and economic royalists. He never wavered from his commitment that the workers and farmers, the "common man" came first. Obama gingerly and gently began to move in that direction in the second year of his administration in 2010. He admitted what everyone knew and that was that making nice with the GOP and making futile appeals to them for bi-partisanship sounded good in White House interviews and congressional speeches, but in the ruthless party-eat-party world of real politick it was a sure fire prescription for an ineffectual, moribund, and hapless presidency, not to mention ridicule as a president without a spine. There was a good reason for the cautious steps Obama took toward FDR.

The GOP was stupendously successful through much of the last century in tagging any Democrat who championed increased regulatory powers, higher taxes on corporations and the rich, greater public spending on health, education and job programs, and bolstering entitlement programs as a reckless, tax-and-spend enemy of private enterprise. Franklin Roosevelt was no exception to the maligning.

There is much that is forgotten in the historic lionizing of FDR for standing government on its head to blunt the hard edge of the Great Depression. The GOP (with some help from a small but pesky clique of Democratic congressional conservatives, big industrialists and conservative newspaper moguls) fought FDR tooth and nail on every one of his reform proposals from Social Security to tighter industry regulation.

Also, forgotten, was that FDR had to tweak, compromise and water down his proposals, even the successful ones, to get passage. His more far-reaching proposal for a national health care plan never got off the ground. No president, and that included FDR, had been a harder target of the GOP attack line on government than Obama.

A vast storehouse of political slurs, snide innuendoes, verbal broadsides and name calling was heaped on his head. The aim was to permanently tag him as the penultimate example of a Democratic president who would make big government the all-embracing, all encompassing, arbiter of American life, at the expense of the private sector. The withering assault by the GOP forced Obama to bend over backward to conciliate, compromise, water down, and even shelve, many proposals to expand government protections and benefits to poor and working persons.

* * * * *

After re-election in 2012, Obama got just enough of a popular mandate to be more daring on Medicare, Medicaid, and Social Security, and dumping the Bush tax cuts for the wealthy. He also proposed a sharp increase in the minimum wage, and spending even more billions on public and private job growth.

This brought back the howls from the GOP of a big government overreach and dire warnings that this would gut big business and thus stifle economic recovery. The comparison to the ire against Roosevelt was in order here. Following his landslide re-election victory in 1936, he ignored the administration baiters and lurched left. He increased spending on job programs, continued to pound the "economic royalists" for subverting the economy, and attacked auto and steel giants and the superrich "Sixty Families" for doing everything to stymie the recovery.

In a fireside chat, FDR talked bluntly with the American people immediately after the 1938 election and made it clear he would not reverse course and that he'd do everything he could to "create an economic upturn" by keeping the government firmly in the business of creating jobs and economic security for the millions still suffering from the Depression.

In 2013, the GOP retained the House, and its party line was unchanged: that millions of Americans still loudly clamoured for a return to fiscal conservatism, and a sprint backward on expanding government programs in education, housing, and highway and urban infrastructure construction and reconstruction. Polls showed the exact opposite.

Most voters wanted Congress to work with the president on solving the nation's problems, most importantly that meant the economy. There was too much political risk in the GOP's mounting an all-out frontal attack on Obama and his plan for more government spending in vital areas. So, the new code word for that was simply to continue to pound on the need for deficit reduction and fiscal restraint. This had just enough public and administration resonance to appear sensible, moderate, and reasonable.

Obama did not take the bait. He gave no sign he'd bow to the GOP's end run around government. He came up with proposals to stave off the hardest and most devastating cuts to everything from Head Start to food nutrition programs and hundreds of other programs for the poor and working class that the GOP demanded. He endorsed enactment of a modified version of the Glass-Steagall Act. That was the tough FDR era bank regulation act. Obama continued to signal that he would plough stimulus dollars directly into government run job training programs, job banks, and public works projects. The other FDR touch was to virtually order the banks to lend more to distressed homeowners, cut borrowing rates and terms, and promise more aggressive government intervention to aid strapped endangered homeowners. These were the programs that would do much to help the working class, and the minority poor.

FDR didn't panic in the face of the GOP's portrait of him as an architect of evil government. He stayed the course, remained true to his populist faith, turned the tables on his foes, and dared them to move the country forward not backward.

Obama appeared willing to do the same. That meant slaying the GOP myth of "evil" big government. Given the massive pressure, pulls and tugs on and at him from many sides, as well as the inherent constraints of governance, he did not totally succeed in slaying the myth. He did, though, succeed in standing fast on the need to maintain the historic role of the federal government as an active advocate and arbiter to provide a high level of government spending, programs, and services to meet the needs of millions of Americans. This was something that private industry could never do. In that sense, he did the best he could within the limits of his power to make like FDR.

CHAPTER 4

Kicking the Foreign War Syndrome

Much was made in October, 2011 by some critics that Obama's announced scheduled Iraq withdrawal was much less than met the eye. Their points of criticism were that 50,000 troops would stay in the country, at a cost of billions more of US tax dollars, and there was no firm guarantee that they'd go packing, one and all, any time soon. The biggest criticism, though, was that Obama fudged on his campaign promise to end the war and remove all troops from the country by May 2010. He made the promise more than two years before that in 2008 during the presidential campaign. When he made it he was careful to note that he meant withdrawal of combat troops. By ending the war, he meant ending American direct involvement in ground action. Many progressives and antiwar Democrats took his "end the war" pledge too literally; the nuances were either ignored, tuned out, or in a euphoric and wildly hopeful moment, distorted.

The Iraq withdrawal, by any measure, was a solid administration accomplishment. The six-year ground war with US troops taking casualties, inflicting death and destruction on towns and villages, and heavy collateral damage, i.e. civilian deaths, stirred international, and regional hatred of the US, and reaffirmed the US image as the bully boy of the world. The war was a colossal domestic and international disaster; and the mountainous lies and deception that Team Bush used to get and keep the US in Iraq will be a permanent mark of historical disgrace and shame on the Bush legacy.

The 50,000 or so troops that would stay in the country for a time, however, was unfortunate, but deemed a necessity. The final withdrawal date of December, 2011 spelled out in the US-Iraqi treaty for the 50,000 could have easily been shaken by any number of events and contingencies. The worst one being a full-blown descent into factional religious or civil war in the country with the US forces caught in the middle. The troops were there supposedly to see that that didn't happen and the training, intelligence and transportation and logistical support, supply for the Iraqi was intended to do just that.

Far more important, though, was the political consequences of the withdrawal. The Iraq war was never simply a military contest to get rid of a hated dictator. In a country that supposedly posed a massive threat to Israel and moderate Arab governments. It was a political war waged to assert American political dominance, control strategic oil resources, to bolster the military hawk credentials of the Bush administration and to boost Bush's tenuous and sagging personal image and popularity on the home front. Obama understood that as long as the bullets, American bullets, flew at Iraqi targets, the US would continue to suffer the deeply flawed and failed political consequences of its overt military involvement in the country.

Obama learned another lesson, a negative one, from Bush's Iraq folly. He did not declare "mission accomplished" as Bush earlier did with the withdrawal. The mission accomplished boast would be tantamount to declaring the war a US victory. To tout a war that should never have been fought and then fought for the wrong reasons, and in the wrong way, would have been laughable and insulting. The crucial point was the death and mayhem Iraqi civilians still suffered there was being wreaked on Iraqis by other Iraqis, not US troops.

The Iraq war was an ugly and shameful page in US history. Obama early on recognized that, and recognized that millions of Americans were furious and frustrated by it, and that the presidential candidate who would do something to change that would be cheered by Americans. Obama deserved cheers. For progressives who cheered him the loudest for pledging to get out of Iraq during the campaign, there was no reason to stop cheering him for fulfilling that pledge in the White House. Obama's Iraq withdrawal was a great deal.

* * * * *

Afghanistan, though, was another matter. In August, 2007 then US Senator Obama fresh on the presidential campaign trail made an impassioned promise at the Woodrow Wilson International Center for Scholars to wage what he dubbed the war that has to be won. The war was the war in Afghanistan. From those early speeches Obama was doggedly convinced that the Afghan war could be won, no matter the cost. He was willing to stake the credibility of his administration on that, no matter the price. The price was high. Polls found that more Americans than ever said the war was pure folly. A majority of Obama's most fervent backers said the same. With the grumbles from liberal Democrats and progressives getting louder about Obama's stance, Afghanistan loomed even larger as Obama and the Democratic Party's Vietnam. Vietnam, of course, was the dreaded word that Presidents Lyndon Johnson, Bill Clinton, and Bush heard about Vietnam, Somalia, and Iraq. It's still the poster war for a failed, flawed, and hopelessly unwinnable war. The word has been a political tipping point for presidents. Vietnam soured public opinion, drained the economy, fueled public dismay and anger, hampered passage of their domestic programs, fractured their party, and stirred big losses in Congress. Public shell shock over unpopular wars always redounds to the advantage of an incumbent challenging a president whose name is linked to the war. In 1952, Eisenhower ran on the pledge to visit Korea if elected. Though Ike never directly promised to bring the troops home if elected, the implicit commitment was that if elected he'd do that. He really didn't have to make that promise; public weariness over the war was so great that Ike's generic oath to visit the troops would have been enough to help sink Democratic President Harry Truman if he had decided to seek re-election. In the public's mind the Korean War had become Truman's war, or, more accurately Truman's failure to win the war. Similarly, Richard Nixon learned from Ike. During the presidential campaign against Democratic Vice-President Hubert Humphrey in 1968, Nixon dropped careful politically calculated hints of a "secret plan" to end the Vietnam War if elected. Like Ike, he didn't spell out in any real detail just what his secret plan was. And like Ike, he didn't really have to. Public revulsion over Vietnam, as in Korea, was so great that even the scintilla of a suggestion that Nixon could end the war aroused voter optimism for him and even greater fury against Humphrey who was widely seen as the caretaker of Johnson's war (Johnson saw the handwriting on the wall and declined to run again in 1968). These two unpopular wars did in Truman and the Democrats in 1952, and President Johnson and the Democrats in 1968. They also had a tsunami effect on Democratic elected officials. In both election years, the Democrats had a decisive edge over the Republicans in Congress, a wide body of public support, and political prestige. Eisenhower, and later, Nixon, painted Korea and Vietnam as a hopeless muddle that Truman and Humphrey (in tandem with Johnson) made a mess of. The two Democratic presidents paid dearly for it, and Bush and the Republicans paid just as dearly for the Iraq quagmire.

Obama knew that history well. He embedded that history into his first presidential campaign and continually reminded voters of the history of the Iraq war failure. Financially draining wars take a huge toll on the economy, drag down public morale, and cause a steep plunge in American prestige internationally. It also whips up greater anti-American sentiment. Three failed and flawed wars and the public's distaste for those wars helped topple two sitting Democratic presidents and hopelessly discredited a Republican president. The same public distaste for the Afghanistan war could easily have made it Obama's Vietnam. History served notice on Obama of this peril.

Unfortunately, Obama never solved the Afghan dilemma. He would leave office with a war still raging, an Afghan regime, riddled with corruption and division, and a Taliban and anti-US opposition forces marking time until the day the US packed up and left. Afghanistan was a tough nut that would remain, if not a taint on Obama's peace and diplomacy legacy, a troubling question mark about it.

* * * * *

The Afghanistan quagmire that remained a quagmire during Obama's White House tenure was not the only foreign policy thorn for him. There was also Syria. GOP congressional leaders spent months in 2014 screaming at Obama to take action against Syria. Their first call was to take out Syrian President Bashar Assad's regime. Their later call was to hit hard at his military's alleged use of chemical weapons against civilians. Then GOP congressional leaders deftly switched gears and demanded that Obama seek congressional approval before launching any strikes against the Assad regime.

If Obama had bowed to the initial GOP demands to take military action before getting congressional approval, GOP leaders would have jumped all over him and blamed him for not getting that approval — and they wouldn't have stopped there. They would have also blamed him for any real or perceived failure to remove the Assad regime, or failure to sufficiently degrade his military to enable the rebels to defeat it, or for igniting a blowback that would further strengthen the hand of al-Qaeda linked radical Islamist factions fighting to topple Assad.

They would have sanctimoniously waved the provision in the Constitution that strictly forbids a president from engaging in any military action in the absence of an actual or imminent danger of an attack against the United States. The president does have authority under the 1973 War Powers Act in some situations to wage limited war with a strict time constraint on it without congressional approval, but even here, at least on paper, the president must report to Congress on the action. The GOP rendered this a largely moot point when it switched gears and hoisted the demand that Obama take no action without congressional approval.

Obama recognized the "damned if I do, damned if I don't" trap that the GOP had laid when he tossed any decision to strike Syria back to Congress. But even this didn't silence Obama's GOP critics. They quickly trotted out the line that Obama wasted time by not consulting Congress in the first place, and that this further contributed to the muddle in Syria. The GOP's shallow, and self-serving gyrations to hammer Obama notwithstanding, the truth was that the type of unspecified intervention that the GOP demanded Obama take, would have had potentially disastrous unintended consequences from killing innocent civilians to further stirring anti- American hatreds in the Middle East, while doing nothing to rid Syria of Assad. The Obama administration would have had to answer for those consequences.

Obama, fortunately, had enough of the history of the Bush administration's bungles in Iraq and Afghanistan on his side to draw lessons from. This started with his own accurate assessment that he made of the Iraq war as a presidential candidate in 2007. He lambasted the Iraq war, calling it the wrong war in the wrong place at the wrong time. Even more pointedly, he called it "a dumb war." Instead, Obama pushed Afghan intervention as the right war, but with the caveat that even with this war, there should be specific goals, specific timetables, and a specific exit plan for getting out of that war.

In his pre-presidential speeches, interviews and comments on the war, Obama massaged his war plan. He promised to set a timetable for eventual withdrawal, get out of Iraq and corral America's European and Middle East allies in a partnership to wipe out the terrorists and their mass destructive weapons. He also vowed to end corruption, hold free elections, bolster Afghan security forces, boost intelligence gathering and monitoring, beef up Afghan security forces, and ensure a stable government in Afghanistan. This was a tall order by any measure. In the years after he spelled out that plan, the United States shelled out between $4 to $6 trillion on the Iraq and Afghanistan wars with no long-term guarantee that the goals of the wars would be met.

The GOP's politically cynical demand that Obama intervene in Syria without fully assessing the gains, and losses of a headlong rush into the country's civil war, and without an accurate scorecard to tell which warring faction in the country would truly bring humane, democratic rule to Syria, would have been foolhardy at best and utterly disastrous at worst. Obama fortunately didn't have to make a one man stand on Syria with all the perils and pitfalls that that would have engendered. Not the least being the trap that the GOP had hoped to set for him. This was one war that could have been. Obama wisely and deftly kept the country and his administration out of.

* * * * *

Wars, of course, are always the most dramatic and painful measure of how effective a leader is in walking the often fine and blurred line between diplomacy and negotiation and the use of force and waging war to attain a nation's foreign policy goals. Iraq, Afghanistan and Syria were the prime examples during Obama's White House tenure. However, there were other gauges of effectiveness. Obama did not have to be defensive about military cuts for which he took some heat. Countless defense experts and those in the defense establishment were absolutely unequivocal that the phased in military reductions in tandem with what the Defense Department had projected did not weaken one whit America's strength and ability to respond to a crisis.

His relations with Russia and the European allies, and most importantly his outreach to the Muslim world were crucial. There were rough edges and tense moments in each of these policy areas. But on balance, his willingness to engage allies and enemies at the table and travel thousands of miles to engage in talks did bring peace, stability and healthy relations with European allies and a cooperative relation with the Russians. He managed the always tenuous feat of keeping the line of dialogue open between the Palestinians and Israelis, without weakening U.S. and Israeli relations. It was his outreach to the Muslim world, though, that stood out. This was the one region that has been the perennial Achilles Heel for immediate presidents. His on-going initiatives to Muslim leaders was the best guarantee of America's defense and security. This was a prime impetus in the deal he ultimately brokered with Iran in 2015 that limited Iran's nuclear program in exchange for relief on economic sanctions. It was controversial, and hotly opposed by many GOP leaders. The ironic proof of the importance of that deal to America's security came in one of the first pronouncements of President-elect Donald Trump. He said he would not cancel the deal. This was the signal that Obama's efforts in war and peace, no matter how many bumps, twists and turns they took, would mostly stand the test of time.

CHAPTER 5

Rough Road to Benghazi

In September, 2012, terrorists attacked the U.S. Mission in Benghazi, Libya. This presented a major crisis for Obama. The terrorists killed four Americans and wounded two others. The crisis wasn't just over the number of American lives lost. It was again the classic case of what did the president know, and when did he know it, and did he do nothing about it, and then try to cover up his inaction? Obama was hit with the charge that he sat and watched the attacks unfold in real time and denied all request for military action during the attacks. Obama was on the hot seat and GOP leaders were turning the heat up even higher. There was a huge political incentive to do that. The attack came just weeks before the November, 2012 presidential election. The fall-out from this could easily have put Obama's re-election in jeopardy.

It was certainly clear from a series of emails between officials that the White House was informed that the attack on the American mission in Libya may have been a planned, organized terrorist attack. Obama's critics jumped on this disclosure supposedly as the smoking gun proof that the White House engaged in a massive cover-up to keep the truth from the American people. But what was the truth? And what proof was there that Obama had conclusive evidence that the attack was the handiwork of a terrorist organization? And just what terrorist organization might that have been? Even if more information about whom and what was behind the attack were known at the time, what motive would Obama have had for covering up the information?

* * * * *

There were no clear answers to any of these questions then. The emails of the White House and the State Department that were released later in and of themselves didn't provide the answers to any of these questions. The larger question, though, was would the Benghazi attack do any damage to Obama's reelection drive? The GOP, assorted web sites, bloggers, and the usual anti-Obama critics, had a field day trying to paint the administration as inept and with a huge credibility gap at best and, at worst, deceitful. Some even likened this to Obama's Jimmy Carter moment. That equated the Libyan debacle to the Iran hostage crisis that was generally regarded as a prime reason that Reagan got a boost in the closing days of the 1980 presidential campaign and ultimately led to Carter's defeat.

There was no comparison between the two catastrophes. The Iran hostage crisis stretched over 444 days, and involved a long train of inept, failed diplomatic initiatives, a disastrous military action, all set against a fanatical nation state that was determined to use the crisis to give America a political black eye. There was also the hint of a deal to delay the release of the hostages until after the presidential elections. None of these factors were at play in the Libyan mission disaster.

That didn't mean that Benghazi had no political significance in Obama's close election battle with GOP presidential contender Mitt Romney. It would not be a significant factor. The reason went beyond the issue itself. It struck to the heart of how foreign policy issues had been reduced to the backwater, as almost after thoughts during the presidential campaign. The foreign policy debate was textbook proof. The Libyan mission fiasco was quickly brushed aside, and while there was some talk about Iran, Israel, and China, Obama and Romney sped quickly back to domestic issues, to pound each other on who best would handle the issue of jobs and the economy.

One of the presidential debates between Obama and Romney centered on foreign policy with Benghazi as a center piece issue. It was the least watched of the three debates that they held that year. This further underscored the consensus point that Americans may have regarded Iran's potential nuclear armament with a nervous eye and had questions about Benghazi. However, they were far more nerve wracked by the prospect of being jobless, under-employed, and having even less in their paychecks and pocketbooks.

If 2012 had been any other presidential election, and the potentially dynamite revelation that the incumbent president possibly fumbled the ball on a grave foreign policy disaster such as Benghazi, the challenger would have pounced on it with both feet. But Romney was virtually mute much of the time on what Obama did or didn't do about the embassy killings. He continued to hammer away on his one and only theme that he, not Obama, could best turn the economy around.

Even if Romney had chosen to take Obama to task, given the questionable circumstances about the White House's knowledge of the Benghazi attack, it still wouldn't have had much resonance. It didn't mean that it would go away. Though there wasn't much likelihood it could hurt Obama directly, he was re-elected handily, there was another way to come at him on the issue. That way was through his Secretary of State, Hillary Clinton.

* * * * *

It didn't take long for GOP congresspersons to zero in on Clinton as the villain. In the months and years after Benghazi, she would be dragged before select congressional committees controlled by the GOP to testify about various aspects of the Benghazi episode. In her third appearance before the committee, and three years after she left the post of Secretary of State, the panel honed in on her private emails to make the case that as then Secretary of State she either badly bungled the U.S. response to the Benghazi attack or deliberately lied about what she did or didn't do to prevent the attack.

The emails supposedly contained classified material that could have been crucial to establishing the GOP's case against Clinton. The GOP quickly jumped all over a report from a government inspector general that a handful of Clinton's personal emails contained classified material. The government investigator did not say what the nature of the classified material was, whether the alleged transmitting of this material was inadvertent, or whose responsibility it was to insure the classified material was marked as classified.

That didn't matter. The relentless GOP narrative was that Clinton illegally used her private email account to conduct State Department business and that those emails would show a pattern of neglect, incompetence and cover-up. The story got even bigger when a tantalizing hint was dropped that there could even be a criminal probe into Clinton's supposed misuse of her emails.

None of that was new stuff. It had been hashed and rehashed over so much that if anything the raw facts about the emails were blurred or distorted. The facts, though, were clear. There was absolutely no proof that Clinton did anything wrong in using her private server to conduct some State Department business. There was no legal prohibition on any government employee using their personal email account to conduct government business at the time she did. The GOP tried hard to poison the public well with the suggestion that

Clinton may have destroyed or mishandled, either deliberately or accidentally, government documents tied back into Benghazi.

There was no proof of that. However, with the 2016 presidential campaign now in high gear, it gave the GOP just enough pretext to again try to taint Obama with the scandal. GOP presidential candidate Trump wasted no time. He told a campaign crowd two weeks before the November 8 vote, that Obama knew about Clinton's private email server and the inference was that he also knew about classified material being mishandled by Clinton. Obama's earlier admission that the email flap did pose a national security threat didn't help matters. It further clouded the issue since Clinton worked directly under Obama when she allegedly used her private server to send classified state department materials out.

There was no evidence that any of the emails pertained directly to the Benghazi attack, let alone anything that hinted of laxity in Obama's response to the attack, or an attempt to cover up how he responded to it. Nonetheless, the point of the GOP's Hillary email obsession was to pound her for her alleged missteps and conniving on Benghazi and in the process, embarrass Obama.

The best that could be said about the long drawn out Benghazi drama was that there was never any proof that Obama downplayed the danger to Americans and when the attack occurred tried to cover up an administration fumble. However, the attack did leave a bad taste in the mouths of many Americans. It gave the GOP much ammunition to further hector and harangue Obama. The worst was that Benghazi continued to hang in the political air as an issue for some during Clinton's presidential bid. That's because Benghazi from start to finish was as much about Obama as Clinton.

CHAPTER 6

The bin Laden and Guantanamo Take Down

During the 2008 presidential election campaign, the GOP plan of attack on then-candidate Barack Obama was simple: slam him non-stop as soft on terrorism and antagonistic toward the military. GOP presidents Reagan, Bush Sr., and especially George W. Bush in his 2004 re-election fight against Massachusetts Senator John Kerry, used this ploy masterfully against their Democratic opponents. The GOP strategists believed that the soft-on-terrorism smear would work even better against Obama. He was a liberal Democrat, untested in foreign policy matters; he had made conciliatory remarks about Islam; he was a staunch opponent of the Iraq war, and unstated, but very much a part of the thinking, he was African-American. This supposedly made him vulnerable to the widely held, borderline racist suspicions that blacks are unpatriotic. The smear almost worked. Polls consistently showed that despite the mountain of political baggage carried by GOP presidential contender John McCain, his running mate Sarah Palin, and the rest of the Republican Party—not to mention the sky-high voter disgust with Bush's domestic and foreign policy bumbles—the terrorism issue still had enough resonance to keep McCain competitive. Obama knew the history of how the GOP used the terrorism ploy to discredit Democrats, and he moved quickly to counter their attacks. He threatened preemptive strikes against Pakistan for harboring terrorists and vowed to wage fierce war in Afghanistan against terrorism and Al Qaeda. During the campaign, he promised to launch preemptive strikes against terrorists wherever they were, including search-and-destroy missions to ferret out bin Laden. He even quipped that he'd put his own life on the line to stop another 9/11. To the GOP's shock—to the ire of many progressive and liberal Democrats—Obama was as good as his word. He refused to soften any of the provisions of the Patriot Act, promptly issued a shoot-to-kill order against Somali pirates who had seized American hostages, stepped up the drone attacks on the Taliban in Pakistan, and approved the massive expansion of troops, bases, and spending on the Afghan war. Most importantly, he issued tough (and secret) orders to the CIA to continue to do everything to destroy and disrupt Al Qaeda and to take out the one man whom Americans most wanted dead: Osama bin Laden. Obama's order to the CIA and military counter-terror teams hunting bin Laden was clear—Do not capture, but kill. The sudden, stunning success of that mission in hunting down and killing bin Laden in May, 2011, forced GOP leaders to scramble. The cheering crowds outside the White House following the announcement that bin Laden was dead, the glowing praise from much of the public, and the congratulations from world leaders drove home the frightening political implications for the Republicans, just as the 2012 campaign was gearing up. Obama had done the one thing that Bush, despite his bluster and tough talk, could not do— take out America's Public Enemy Number One, the world's preeminent symbol of terror. In one fell, and spectacular, swoop, Obama rudely shattered the myth that had been a key weapon in the GOP arsenal for decades: that a Democratic president was incapable of waging a tough and effective war on terrorism. Confronted with the political game-changer of the bin Laden killing, it was amusing to see the tortured gyrations that conservatives and GOP officials went through to heap credit on Bush, the military, special ops teams, the CIA, the 9/11 victims' families, and even the general public, all to avoid giving the credit where it properly belonged. That was to credit Obama for giving the order to hunt him down.

* * * * *

The bin Laden action came at the worst possible time for the GOP that year. Obama's approval numbers had been sagging, and more Americans continued to voice displeasure over the direction the country was heading. This was the one bright spot for Republicans, especially coming on the heels of other polls that showed the GOP base was bored, disheartened, and even contemptuous of the crop of would-be 2012 Republican presidential candidates. At the same time, many voters were repelled at the clownish showboating of purported GOP contenders Donald Trump, Sarah Palin and Michele Bachmann. In his televised address announcing bin Laden's death, Obama wisely did not revel in this victory. He portrayed it solely as a grim but necessary action in the war on terrorism. Bagging bin Laden was purely a national security priority. Just as smartly, the president took pains to assure that the attack on Al Qaeda's leader was not an attack on Islam. Both messages were necessary, and both left the GOP even more hapless and reeling for a brief time for a way to recover politically. The take down of bin Laden shattered another of their cherished myths about Obama being a soft touch on terrorism.

At the same time this did little to quell the clamor for Obama to punctuate every utterance he made about the threat of ISIS or Al-Qaeda terrorist attacks with the words "Islamic terrorism." For instance, the moment news broke about the hideous terrorist attack bloodbath in Paris in November 2015 that resulted in scores killed, the predictable happened. The tweets from the conservative echo chamber came fast and furious. They called for a total, all-out war, bashed Obama for refusing to bash Muslims, all Muslims, for the acts of a terror driven few.

Obama denounced the attack, and again called for greater vigilance and aggressiveness against terrorism. But he did not say "Islamic terrorism." His care in putting the blame and the action needed in the right place, and not on all Muslims or immigrants was the type of caution and restraint needed at a time when the witch-hunt against Muslims was again escalating.

The right-wing's manic desire to turn every horrific and bloody terror attack into a political vendetta against immigrants and Muslims was a worn script. But it was a serviceable one. In part because there were just enough real, legitimate and bloody terror cells and individuals who were willing to commit mass murder and mayhem for their warped, crackpot, ideals. The other part of the reason the right jumped on the anti-Muslim bandwagon when terror strikes was because it stoked, fanned, and inflamed anti-immigrant fear and hysteria. This has been a trump card for legions of GOP demagogic, fear-mongering officials and candidates.

The Paris attack by alleged ISIS affiliated terrorists in November, 2015 put much pressure on Obama to take an even harder line on the Middle East conflicts. However, it would not have stopped there. It would not have been just ISIS that would have been the focal point of any stepped up American offensives. Muslim groups in this country would draw even greater scrutiny up to and including widening surveillance and monitoring.

This was much in evidence two years earlier following the Boston Marathon bombing in 2013 perpetrated by two muddled, and disaffected brothers who were Muslims. Public anger and passions were running hot. Muslims were again under intense fire, and the clamor was loud for a big crackdown. Obama did not overreact. He damped down the furor and again made clear that the terror attack was the work of individuals and they would be the ones hunted down. Once they were apprehended, their civil liberties and legal due process would be fully protected. He was again careful not to give any hint that the heinous act of 2 individuals had anything to do with Islam or the sentiment of Muslims in this country. He, in effect, played a reasoned, responsible, and decisive role in heading off any anti-immigrant or anti-Muslim backlash to the violent acts of a few.

The Paris massacre was shocking and horrific. It, as all terror attacks did, caused untold personal pain and suffering. However, Obama recognized that the overwhelming majority of Muslims from war-torn countries worked hard to rebuild their shattered lives. They have been among the first to loudly denounce the despicable acts of those who sully Islam by committing abominable acts. Obama backed them when he sent the strong message to hunt terrorists such as bin Laden, not witch-hunt Muslims.

* * * * *

The take down of bin Laden in the larger scheme of the on-going war on terrorism was a simple proposition in comparison to the much greater, and more contentious issue of how to take down Guantanamo. In February, 2016, with less than a year left before he exited the White House, a plainly frustrated Obama in a near twenty minute address was blunt,

"I am very clear eyed about the hurdles to finally closing Guantánamo: The politics of this are tough. I don't want to pass this problem on to the next president, whoever it is. And if, as a nation, we don't deal with this now, when will we deal with it?" Guantanamo, opened in 2002, was the brainchild of George W. Bush as the ideal place to send the detainees from the Afghanistan war. Bush made it clear that they were not protected by the laws of the Geneva Convention or any civilian courts and therefore the government had carte blanche to do with them as it pleased. The countless reports and protests over torture, including waterboarding and other outrageous gross violations of civil rights, was justified under the guise that these individuals were genuine threats to commit terror acts against the U.S. and were deemed outside established legal protections.

In doubling back again on what to do about Guantanamo, Obama almost certainly knew that his last proposal to close the prison which entailed transferring the remaining three dozen prisoners held there to federal prisons in the U.S. had absolutely no chance of being approved by Congress. Both congressional Republicans and some Democrats were vehemently opposed to his plan. In fact, Congress had a law in place that banned the military from transferring detainees from Guantánamo to the U.S. There was almost no hope that Obama's successor, Trump, whose only comment about Guantanamo during the campaign was to say that there were "terrible people" would try to buck Congress on the prison.

This was simply a repeat of the frustrating blocks in

Obama's way from the beginning of his presidency in 2009. He had promised during the campaign that if elected he would close Guantanamo. He repeatedly called it an embarrassment and a flat out violation of international law to hold dozens of suspects there under the mere suspicion of being eminent terrorist threats to the U.S. Obama made one effort to by skirt around Congress with an executive order his first year in office that directed that the prison be shuttered within a year. The political support was paper thin for this and soon evaporated. The failure left Obama wide open to the charge from progressives and many liberal Democrats that he had reneged on his promise to close the prison. Guantanamo was suddenly a lose-lose proposition for Obama. He couldn't budge Congress. And he couldn't convince his left critics that he couldn't accomplish the goal.

The best that he could was peck away at the numbers of prisoners being dumped there. Some were put on a parole or transfer list to further whittle away the numbers. However, whether there was one or 100 prisoners there, the prison remained open. Again, there was nothing that Obama could do about that in the face of a hostile Congress.

He closed his February, 2016, talk with an admonition which he had made countless times before, "Let us go ahead and close this chapter." But it was a chapter during his Oval Office tenure that would remain open. The question dangled, would Obama be remembered as the president who made a real effort to right the outlandish and shameful abuse of international law and human rights that Guantanamo represented? Or would he be reviled as the president who vowed he would close the prison and eight years after he made the vow he failed to keep? The truth is he'd be reviled by those who demanded and believed that he could close the prison and didn't. At the same time, he'd be praised by others who would give him credit for trying to close it. For Obama, though, the frustration over his inability to close Guantanamo might be the one issue that would remain a testament to a president's inability to act long after he left the White House.

CHAPTER 7

Cuba Libre

When President Obama touched down in March, 2016, at Jose Marti Airport outside Havana, he made both history and controversy. The history part was easy. He became the first US president to visit Cuba in nearly a century and more significantly, the first US president, to visit the country since Fidel Castro took power in 1959. The controversy part was a bit more clouded. He was roundly attacked for the visit and his push to normalize relations with Cuba by the usual coterie of suspects. They included older, conservative, Cuban nationals in the US, right-leaning evangelical groups, and, of course, some conservative Democrats and much of the GOP.

New Jersey Democrat Bob Menendez, led the charge, "When we see a photograph of the President of the United States laughing and shaking hands with the only dictatorship in the western Hemisphere, I will be thinking of Berta Soler of the Ladies in White and her fellow human rights and democracy advocates," He referred to a Cuban dissident leader and her followers. A decade earlier any mention of a president making any overture to Communist Cuba would have been unthinkable, and would have been tantamount to political suicide. However, times had changed for many, and that included younger Cuban-Americans who had no personal experience with living in Cuba before Castro.

It also included many leading American corporations, trade and tourist groups that salivated at the prospect of doing business with the island. Florida Senator Marco Rubio, former Arkansas Governor Mike Huckabee, former Pennsylvania Senator Rick Santorum, Texas Senator Ted Cruz, and former Florida Governor Jeb Bush, also repeatedly chimed in and spoke out against any normalization of relations with Cuba. Rubio was even more strident on the pending thaw, calling it "disgraceful." That had to be taken with a whole bag of salt since at the time all had their eye on a 2016 White House bid. As in the past, they were playing the anti-Castro card to the conservative GOP base.

For his part, Trump, the eventual presidential winner, sent off all kinds of mixed signals about what he'd do about US—Cuban relations. At one point during the campaign, he denounced and deplored normalization. Another time he said he'd take a wait and see stance, but insist on a better deal for the US.

So, when Obama told a throng of staff persons at the US Embassy at a Havana hotel that his visit was an "historic visit," and that 'it was an historic opportunity to engage with the Cuban people," he was on firmer ground than any other past president. In the days, after his visit, he got mostly high marks for the visit and lifting the decades of freeze in relations between the two countries.

There were two big takeaways from Obama's Cuban opening. The first was obvious. After more than a half century of U.S.-backed invasions, covert efforts to sabotage and overthrow Castro, an embargo, and a Cold War freeze in diplomatic relations, the U.S. policy toward Cuba was an abject failure. Fidel's brother, Raul Castro, remained the official government head, and Fidel, was still a presence in Cuban life and a bigger than ever figure internationally. Obama took the logical step that almost certainly would have been taken years ago, except for a politically retrograde GOP and older, politically connected Cuban Americans, and that was to normalize relations with the island. Obama pointed to the obvious when he said the old policies, meaning containment and subversion, didn't "make sense." More Cubans were travelling to wherever they could get a visa, political dissent and expression was more open than ever. In fact, Obama met with a range of Cuban dissidents in a confab well-covered by the Cuban press. There were also more private owned businesses and farms in Cuba than ever. While Cuba was still officially a one party-state, Cuban leaders repeatedly made it clear they were committed to real reforms. During the years of the US freeze on Cuba, tens of thousands of Americans had traveled to Cuba. The evidence even then was everywhere of changes in tourism, trade, and people-friendly relations in Havana and other cities.

* * * * *

Given the profound changes in the country, and the polls that showed that a majority of Americans wanted an end to the embargo, Obama's move was more a pragmatic than a bold step. Still, the devil was in the details about how quickly there would be full official diplomatic relations, free trade and free exchange of goods, services and technology, a formal lifting of the embargo, foreign investment, travel, and family relations restored between Cubans in the island and those living here. This would hold true even after the US formally exchanged ambassadors with Cuba and set up its embassy there shortly after Obama's visit. The common sense move to normalize relations was less important than the timing of the move and the domestic political consequences of it. The prolonged and outdated battering of Cuba was absurd because it never posed any real military or economic threat to the U.S. It was about U.S. domestic politics. Ten presidents before Obama were held hostage to the GOP-Cuban lobby and the fear of being branded soft on Cuba. This was tossed at any president and seen as the political death knell for Democratic presidential contenders. Obama's Cuba initiative couldn't be separated from his escalating defiance of the GOP. In the aftermath of its November 2014 mid-term election shellacking, the Democratic

Party was in a desperate search to find its legs. It had been denounced for not fighting back harder on issues from opposition to the Keystone pipeline, the drum beat GOP assaults on the Affordable Care Act and a budget deal in 2015 that was stuffed with financial giveaway goodies to Wall Street. With the White House and some crucial Senate and congressional seats on the line in 2016, Obama was still the key to Democratic hopes for a strong comeback. His willingness to weld the executive pen on immigration reform and a defiant promise to use it whenever and wherever he could to push initiatives that a GOP-controlled House had stymied at every turn was vital to the party. Each of the eventual 2016 Democratic presidential contenders, Hillary Clinton, Martin O'Malley and Bernie Sanders were quick to pick up on the significance of Obama's Cuba normalization proposals and applaud them. It put the Democratic Party firmly on record as reversing a dead-end policy that had been an albatross around its neck for decades. Clinton, who got the nomination, became the first of the official Democratic presidential candidates to call for full normalization. Obama's Cuba move could not be considered of the groundbreaking magnitude of Nixon's China opening or Reagan's working both sides of the street with the Soviet Union, promoting exchanges between students, scientists, artists, and local officials while proclaiming the avowed intent to bring down the "evil empire" and eventually its collapse. However, it sent a welcome signal that on a long standing sore point foreign policy issue such as Cuba, Obama would not succumb to GOP mania and intimidation. This made his Cuba opening more than just about Cuba.

CHAPTER 8

A Presidential Conversion on Gay Marriage

"What I believe is that marriage is between a man and a woman ... What I believe, in my faith, is that a man and a woman, when they get married, are performing something before God, and it's not simply the two persons who are meeting." Then Illinois state senator Barack Obama was adamant on this point when he sat down for an interview with WTTW Chicago public television in October, 2004. Six years later, and well into his second year in the White House, Obama returned to the subject of gay marriage. There was no change, " I have been to this point unwilling to sign on to same-sex marriage primarily because of my understandings of the traditional definitions of marriage." Through much of that year and well into his third year in the White House, in 2011, Obama was still as he delicately put it "working through" his thoughts about the issue.

Obama's deep soul search on the issue came to a dramatic end on national television in May 2012, when he told *ABC's Good Morning America* host Robin Roberts, "at a certain point I've just concluded that for me personally it is important for me to go ahead and affirm that I think same sex couples should be able to get married." Gay rights groups that had hectored, pushed, prodded, and reviled him for opposing gay marriage now loudly applauded. It was the first time ever that a sitting president had explicitly put his personal stamp of approval on same sex marriage. This was hailed as the long-awaited breakthrough needed in the highest of all official circles to finally get a wary, hesitant, but increasingly open Supreme Court to finally say it's OK to make gay marriage both a right and the law of the land.

Most now agreed that no matter which way the Supreme Court ultimately ruled on gay marriage, pro or con, narrowly or expansively, gay rights organizations would owe Obama a debt of gratitude for pushing the nation's opinion envelope forward in support of gay marriage. It was true that gay rights groups had waged a tireless struggle for a decade before Obama's conversion on the issue against the wall of bigotry, ignorance, religious distortion and intolerance and the towering legal barriers to gay marriage. It was equally true that one of those who stood fast behind one of those barriers was Obama. His self-admitted soul searching journey from passive opponent of gay marriage to having profound conflicts about it, to finally whole hearted support of the right, was well known. What was less well known was that Obama's final epiphany on gay marriage was not just one man's conversion to a controversial cause. The politics of that conversion was profound in that it put the White House imprimatur on it. It was part myth that presidents simply are solely the bellwether of public opinion, and don't make a move without first sticking their finger in the wind to see which way public opinion is blowing. Studies have shown that it cuts both ways. Presidents do have the awesome power to influence, sway, and even change public opinion on controversial issues. There was Franklin Roosevelt's slow chipping away at America's isolationist resistance to the fight against fascism in the run up to World War II. There was Truman's bold stand on integration of the armed forces, and Eisenhower's grudging, but, eventually tough, stance enforcing school integration in Little Rock. There was Kennedy's public face down of the Soviets in the Cuban missile crisis, Johnson's push on civil rights and poverty and Nixon's opening to China. Then there was Carter's Middle East peace initiative, Reagan's public assault on government programs, Clinton's revamp of welfare, and Bush's sell of the Iraq war. These were pivotal issues where a president was able to sharply move the dial on public opinion. Obama's breakthrough stance on gay marriage would be viewed in the same light. Before the president's announcement in 2012, support for same sex marriage had barely nudged toward the halfway mark among those who said they supported it. This was a significant jump from six years earlier in 2006 when support for gay marriage stood at under 40 percent. Even more dramatic was the sharp turnaround in black attitudes toward gay marriage. Blacks had been among the staunchest opponents of gay marriage and were the final trump card that Christian evangelicals and GOP ultra-conservatives played hard to spearhead ballot initiatives in several states banning same sex marriage. On the eve of the president's announcement of his support for gay marriage, black support for it had climbed to nearly 60 percent. Following the president's announcement, all the major civil rights organizations, and even a number of religious leaders, publicly declared their support of gay rights. Many Hispanic organizations also publicly endorsed same sex marriage.

* * * * *

One key in just how a president's view on a controversial issue ultimately matters is whether a president will back up his words of support (or opposition) with aggressive action. This means translating his support into public policy initiatives. Obama set the stage for that over the years even when he opposed or was ambivalent about gay marriage He backed gay rights in speeches and legislation 18 times before he grabbed the White House. He showed the same support and sensitivity in his appointments. He was only the second president to speak at the annual dinner of the Human Rights Campaign in 2011. The fact is that the group thought enough of his gay rights advocacy to invite him. Obama pointedly lent his name to opponents of Proposition

8, passed as an initiative by California voters in November, 2008, banned same-sex marriage in California to use in mailers that they circulated. He did the same in the case of other similar anti-gay marriage initiatives in other states. A federal court scrapped the Proposition in a ruling in 2010. The Supreme Court upheld the lower court ruling three years later. Obama sent a subtle signal of his gay rights support on the issue on the Defense of Marriage Act. He could have kept hands off the issue by letting the legal challenge to it run its course. Other presidents had done that when they thought a law was unconstitutional or unjust. This argument though ignored what Obama had said about traditional marriage too, not to mention that he made it plain that he wanted the law repealed—but repealed through legislation that he would push for. One final tip of the decisive importance of Obama's gay marriage support came during the 2012 presidential election. Polls found that the issue would not be a factor in voter support or opposition to him. Even more importantly, the same polls showed that Obama got a bump up in the number of voters who were more likely to back him or at least back him more enthusiastically because of his gay marriage support. Presidents can and often do make a difference, a big difference, in pushing, cajoling, and whipsawing the nation forward on an issue. Obama's stance on gay marriage was proof of that, something gays, and indeed the nation, would for years to come thank him for.

CHAPTER 9

Obama's Border War

In April, 2010, Obama wasted no time in denouncing Arizona's hard-nosed anti-immigration law. The law compelled all immigrants in the state to carry alien registration documents at all times. If they were stopped they had to show the documents, and, worse, it gave police the authority to stop and question anyone whom they suspected was in the country illegally.

He called it misguided, irresponsible and a threat to civil liberties. Obama was right. The bill was wasteful, unenforceable, and more ominously, virtually a license for police to engage in racial profiling. But it was also popular in Arizona, at least from the polls, and beyond the polls it almost certainly reflected the sentiment of a lot of those in Arizona and the jaundiced view of millions more Americans on immigrants in other states.

Arizona officials claimed they had to act in large part because the federal government had dithered, stalled, and back-pedaled countless times on enacting comprehensive immigration reform. That, in effect, dumped the immigration reform issue squarely back in Obama's lap. The president was under withering pressure from immigration reform leaders, Hispanic activist groups, and the Congressional Hispanic Caucus to get him to get the ball rolling on a reform bill in Congress. The demand then couldn't have come at a worse time for Obama.

The loss of thousands of jobs, with official unemployment, at that time nudging double digit, African-American joblessness far higher, and with low wage American workers bearing the brunt of the downturn, presented a wedge for immigration foes. They hammered hard that undocumented workers snatched jobs from needy American workers. The charge was hotly disputed, and there was always much evidence that this was not the case. Yet, the issue and the feeling still touched a raw nerve.

There was also the loose network of anti-immigration organizations and the legions of right wing talk jocks, Tea Party activists, and Fox News Network talking heads who stirred the troops to oppose any reform. The stock attack charge that any immigration reform bill was a de facto reward for breaking the law still ignites anger and passion in many Americans. Then Arizona Governor Jan Brewer tied her signing the bill into law into another issue that ignited even greater passion and anger. And that was crime. She flatly said that the law would help protect her state from crime from Mexico. The governor cited no evidence to show that immigration has bumped the state's crime rate up. But then again, she didn't have to. The frightening shots of bullet riddled, hacked up bodies that have become regular news features on American TV screens from the low intensity warfare in Mexico between government forces and the drug cartels were more than enough to stir terror in many Americans that a wave of illegal immigrants flooding the country would turn America's streets into blood drenched streets.

Immigration reform also can't be separated from partisan politics. With November 2010 mid-term elections only months away there was the real danger that Democrats could lose by picking a fight that was bound to be even more contentious and divisive than the health care battle was just too great a risk.

* * * * *

This put Obama in a delicate spot. He had a major fight on his hands to get a financial reform bill passed. There was the risk that the concessions he and Senate Democrats made to Republicans to quickly get the bill passed could alienate many liberal and progressive Democrats who wanted to see the toughest possible consumer protections in place against the ravages of big banks and financial houses. They were the driving force behind his election win and the White House banked on their numbers and passion to help blunt the momentum of Tea Party activists in the fall, and beyond.

He was walking a tighter line on the issue.

So there was little surprise when Obama gave immigration reform short shrift in his State of the Union Speech in February, 2010. This rankled immigration reform backers. They loudly protested that the president had reneged on his promise to them to make comprehensive immigration reform a centerpiece of his agenda. In the months afterwards they increased the pressure on him to make good on the promise with the vague hint that if he didn't, more than a few Latino voters would be tempted to stay home in the fall and beyond. The Arizona law and Obama's response to it did two things. It showed that the issue was emotional, volatile, and loaded with political peril for the Democrats. It also showed that no matter how vigorously Obama denounced a bad, and repressive immigration law, it wouldn't be enough for immigration reform activists. He'd have to put the full muscle of his office behind a wide-ranging comprehensive immigration reform law. The problem for him was that there was no way this law could ever get through a GOP controlled House. Obama knew that and there was little chance that he would expend the full political currency of his office on this. This put him in a no-win spot and left a bad taste in the mouths of many immigration activists. Two years later in July 2012, the Supreme Court struck down most of the provisions in the law. It was too late to rescue the Democrats in Congress. By then the GOP had tightened its control on Congress. Some blamed Obama for this. It was just more proof of the rancor the issue caused.

* * * * *

It didn't take long for that rancor to resurface. In November, 2014, Obama almost certainly saw and heard the shouts of several dozen protestors who stood across from Del Sol High School in Las Vegas where he made a major speech touting immigration reform. The protestors were relatively few in number. A few years earlier, they would have set off warning bells that immigration reform was one of the most touchy, volatile, and polarizing issues on the nation's political table. GOP leaders had twisted it as a powerful wedge issue to tar Democrats that supported immigration reform, and especially its cornerstone, the path to citizenship, as coddling lawbreakers, and driving down the wage and living standards of American workers. The bitter reality was that polls and surveys repeatedly showed that a majority of Americans across all lines agreed with the GOP hardliners, that immigration reform smacked of a giant reward to those who broke the law and were in the country illegally.

That feeling was so intense that many Democrats ran from immigration reform like the plague. The one Democrat, though, that Latino activists and immigration reform advocates continued to expect action from was Obama. For much of his first term, the word from the White House on immigration was no word. It cost Obama. His popularity among Latinos dipped for a time. That was due almost exclusively to the sense that the president had done next to nothing on immigration reform. Yet, he still desperately needed Latino votes to tip the White House back to him in 2012.

Obama, however, even as his popularity numbers slightly fell among Latinos did not totally ignore the issue. He lashed the GOP for torpedoing comprehensive immigration reform legislation in Congress on the two occasions when it appeared that an immigration bill might be reintroduced.

Obama was not to blame that this didn't happen. The crushing problems and bruising fights over deficit reduction, spending, health care reform, coupled with high soaring gas prices and the jobless crisis were endless and time consuming. The fights required every bit of his political capital and arm twisting to make any headway against an obstructionist, intransigent, and petty GOP determined to make him pay a steep political price for every inch of legislative ground he sought to gain.

The 2012 election changed only one thing with the GOP. That was its in-your-face, xenophobic rants against illegals supposedly stealing jobs from Americans and breaking the law. GOP leaders had no choice but to tamp down their saber rattle immigration rhetoric for the simple fact that Latino voters punished the party mightily in 2012 for that rhetoric, and sent an even stronger signal that it would continue to punish the GOP if it didn't change at least its tone on immigration. The 2012 election changed one other thing. It gave

Obama the long sought and awaited opening he needed to go full throttle on immigration reform.

* * * * *

The election result that year was not the only strong point for Obama on reform. In 2007, then President George W. Bush was widely and unfairly blamed for making a mess of the immigration reform fight in Congress by not pushing hard enough for passage of the bill. Immigrant rights groups lambasted Republican senators for piling crippling demands for tight amnesty, citizenship and border security provisions in the bill. Leading Republican presidential contenders didn't help matters by flatly opposing the bill as much too soft on amnesty and border enforcement.

This did much to kill whatever flickering hope there was for the bill's passage. This undid the inroads that Bush made in the 2000 and 2004 presidential elections when he scored big with Latino voters. A big part of that then was due to the perception (and reality) that Bush would push hard for immigration reform. The GOP didn't learn a thing from this. It was almost as if Bush's Latino vote ramp up was an aberration. The GOP's metallic ear on immigration culminated in the quip from 2012 GOP presidential loser Mitt Romney that the best way to solve the immigration crisis was for undocumented workers to "self-deport."

Obama's battle for the Latino vote in 2012 was never intended to head off any mass defection of Latino voters to the GOP. There was never any chance of that. The polls that showed Latinos less than enthusiastic about Obama also showed absolutely no enthusiasm for any GOP would-be presidential candidate, let alone that there would be a massive vote for GOP candidates.

Obama's frontal challenge to the GOP to do something about immigration reform was not only a long overdue move to right a long simmering policy wrong, but a move that if handled right could have done much to shove the wrenching issue of what to do about the nation's millions that are here without papers, and are here to stay, off the nation's political table.

This was not to be that year or any other year after that. The only option that Obama had was to issue a series of executive orders that pecked at the edges of reform, by refusing the deportation of some undocumented workers, but this was no substitute for a full congressional approved comprehensive immigration reform law. Obama would leave the White House in 2017 with no immigration reform bill to show for his efforts, a tightly controlled GOP Congress showing no sign that it would in any way consider a reform measure, and worse, President Trump vowing to keep his campaign pledge to deport millions of immigrants from the country, and build an impregnable barrier along the border to keep them out.

This insured that immigration reform would not be among the legacy accomplishments of Obama. He'd be known as just one more in the line of presidents, both Democrats and Republicans, who could not get anything done on the much needed major overhaul of America's antique immigration laws.

CHAPTER 10

Trying to Outgun the NRA

The National Rifle Association and key GOP congressional leader's response to President Obama's series of unilateral executive orders on gun control in January 2016 was predictable. The NRA thundered that they were useless, unconstitutional, and wouldn't do a thing to keep guns out of the hands of criminals. The response from GOP congressional leaders was just as predictable: stone silence.

Obama's executive orders that bolstered the nearly two dozen executive orders he previously issued in January, 2015 zeroed in on tougher background checks and more stringency in restricting the reimporting of guns back into the U.S. from abroad. The background check requirement was the deal breaker in March, 2013 when the Senate slammed the door on Obama's and the Senate Democratic leadership's gun control proposals supported by a handful of key Republicans.

The new background requirement was not universal. It was limited to choking off the use of trusts and/or corporations as fronts to buy guns from while evading any background checks. This was not insignificant. The ATF got nearly 40,000 requests in 2015 for transfers of guns to trusts or corporations. How many of these weapons ended up in the hands of criminals, or were used in gun related violence was anyone's guess in the absence of tighter checks over who the corporations or trusts sell the guns to.

Obama took the action as he did in January, 2016 for two reasons. One was the spectacular and grotesque mass killings in Aurora, Colorado and Newtown, Connecticut during his second term. There was the equally heart wrenching murder in Duncan, Oklahoma, the continuing murder plague in Chicago, and other high profile senseless multiple gun related killings. The second reason was just as compelling. Congress was due back in session in September, 2016 and by plopping a tougher background check requirement back on the political table, Obama was in effect laying down the gauntlet again to Congress and by extension the NRA to act.

This was crucial. The new executive orders, as the older ones, did not strike to the heart of how to stop a future mass killing. That was still Congress's job. The key congressional leaders that pushed for more stringent gun control provisions had not totally thrown in the towel. The problem though was how to get conservative Democrats to back those provisions. That job was even tougher given that several of them had broken ranks with the Democratic leadership and opposed the

Senate gun control provisions in 2012. They faced strong re-election fights in 2014.

Still, polls continued to show that many Americans supported stronger bans. In the decade since the expiration of the assault weapons ban, Congress had ample chance to enact a slew of gun control curbs that would have compelled legal gun owners to store guns in a tamper-proof safe and secure enclosure, bar anyone undergoing or referred to treatment for mental or emotional disorders from purchasing or having direct access to guns, and clamping an absolute ban on the sale of assault weapons. These and other measures were repeatedly introduced in Congress and just as repeatedly buried before they ever reached the House or Senate floor.

Obama's executive orders touched on many of the measures that the dogged handful of congressional gun control proponents had pushed for in the past. But they didn't have anywhere near the wallop that congressional gun control laws would pack. Obama was in effect bumping up not against Congress in trying to pry some kind of gun control provision passed, but the NRA. It knew that the name of the game on gun control is Congress. The NRA had been wildly successful in sending the message to Congress to round-file all gun control proposals for years through its well-oiled, well-versed, labyrinth of PACs, lobbyists, legal counsels, divisions, funds, and a foundation. The NRA has these divisions: Federal Affairs, Public Affairs, Finance, Research & Information, Conservation, Wildlife & Natural Resources and, most importantly, the NRA Political Victory Fund. Its scorecard of wins in congressional elections was nothing short of phenomenal.

* * * * *

In 2008 it was directly or indirectly involved in nearly 300 campaigns for the House and Senate. NRA-backed candidates or incumbents won 230 of them. It spared little expense in padding its congressional win scorecard. It ranked in the top tier of contributions received, lobbying dollars spent and money garnered and spent by its PACs. It was not just the NRA's money and willingness to spend it to pack Congress with pro-gun backers. It also cherry picked former government officials or job holders to do its congressional arm twisting for it. In 2012, fifteen out of its nearly 30 lobbyists had government ties. The year 2016 was no different. The gun rights group dumped more than $50 million into a half dozen Senate races; the money went for huge ad buys. All the candidates they backed were Republicans. Only one of the candidates lost his race. During every election cycle that Obama was in the White House the NRA shelled out one million dollars or more toward 14 congressional races. It got its money's worth by winning 11 of those contests. In the 2014 mid-term election, it was the driving force in insuring that the Republicans took back the Senate. It spent more than twenty million dollars alone on the five races that it believed absolutely must win in the Senate and won every one of them.

The assumption that the NRA is basically a front for conservative GOP business and political interests was another faulty misread. Though a big share of the NRA's campaign dollars went to Republicans, it was adept at spreading the largess around. In 2012, some Democrats received more than $2 million in NRA campaign contributions.

The NRA has gotten a stupendous return on the seventeen million it spent on federal elections in 2012 and the tens of millions it spent on past elections. In the decade since the assault gun ban expired in 2004, nearly 20 strong gun control bills have died still born in House and Senate committees. There hasn't been much movement in the states either to get tougher gun control laws. Thirty-three of the states have the barest minimal gun checks. A dozen others have only slightly more restrictive controls on guns.

The NRA has been able to lobby, harangue, threaten, and intimidate gun control advocates in Congress, in addition to defeating pro-gun control candidates. It played watchdog over judicial appointments and moved quick to torpedo the confirmation of any judicial appointee who gave even the vaguest hint that he or she would be likely to issue a ruling upholding a gun curb. This strategy had also paid big paid dividends. Time and again, state and federal appeals courts and the Supreme Court overturned city and state gun control laws.

Obama's executive orders thrust him squarely back into the center of the gun control battle. That amounted to a frontal challenge to Congress to seriously debate the need for gun control legislation. Every round in that debate would be just as fierce, contentious, and the NRA vehemently argued that gun control bans don't reduce gun violence. The NRA also threatened vulnerable Democrats and wayward Republicans that they'd target them for defeat by dumping bushels of campaign money into their opponent's kitty if they didn't tow the NRA line. There was no turning a blind eye to the gun carnage that tore the nation. Obama's executive orders on gun control were a strong reminder to the nation of the needless bloodshed and the refusal of Congress to do anything about it.

Unfortunately, his orders did not, and could not, the change the iron cast gun culture in the nation and the lock down suffocating control of Congress by the NRA. This was not Obama's failure. It was society's and the nation's political leader's failure. That failure was something Obama could not escape from an would dog his White House tenure.

CHAPTER 11

Thunder from the Left

In an interview four months before the end of the 2008 presidential campaign that November, leftist gadfly Ralph Nader was plainly peeved at then Democratic presidential candidate Obama. He lambasted Obama for being what he considered a bought and paid for Washington Beltway insider. He punctuated the rip at Obama by asking rhetorically "Is it because he wants to talk white." A year after Obama's 2012 election win, Nader was back on the attack. He topped the "talking white" dig by calling Obama "a con man" and flatly said that he had no use for him.

Nader and legions of other progressives fervently believed there was no substantive difference between the Democrats and Republicans. That first and foremost included Obama. They didn't buy his campaign mantra that he was a real change guy, but as Nader branded him, just another deal making Beltway shill who had betrayed his hope and change promise. That was the constant refrain from left side Democrats and progressives who pounded Obama for what they believed were flips, reversals and shifts on crucial policy issues, ranging from health care to the Iraq war.

In 2012, things came to a head with the threat of an active and passive drift of progressives away from a full support of his policies. The threat was fed by their deep questions on everything from the compromises he made on health care reform, to the perception of over catering to Wall Street interests. The hard reality, though, was that Obama needed liberal Democrats and progressives in Congress and in the field to sell his initiatives on immigration reform, jobs and the economy, the looming showdown with the GOP over the budget, and his staff and judicial appointments. Conservative Democrats, bankers, corporate CEOs and lobbyists could hardly be expected to generate the passion, energy and, most importantly, the bodies, to do the political grunt work to back him and to spearhead the tough battle many Democrats would face to keep their seats in the House and Senate in 2014.

This was the Catch 22 for Obama; he needed their support. Yet, progressives within and without Congress screamed at the top of their lungs to remind him of the promises he made on health care which was to back a single payer plan. Even though he backpedaled, they implored him to not to back away from the public option: take off the gloves; play hardnosed politics; nose thumb the GOP; and shove a real health care reform bill through Congress. A hard, uncompromising fight would have done much to reassure distraught progressives that there was still some semblance of the progressive Democrat that many thought they were voting for in 2008.

Obama, though, had the standard liberal reservations on crucial issues that caused bitter feelings of progressives toward him. He was not unmindful that the stimulus bailout he signed off on gave too much away to bad behaving, profligate spending banks and Wall Street greed merchants and that massive tax cuts wouldn't do much to help the poor and the middle class unemployed. Even so, he backed the bailout and tax cuts, albeit modified, to get the support of a handful of congressional Republicans. He opposed the Iraq war, but he calculated that if he pushed for a quick and immediate withdrawal, the military brass and conservatives would have howled and dug in their heels to resist. Instead, he talked about flexible timetables and a phased withdrawal. In the case of Afghanistan, he chose to send the strong signal that he'd rush even more troops into the quagmire. That assured that just as Iraq was seen and pilloried as Bush's war and folly, Afghanistan would be seen and pilloried as Obama's war.

Liberal Democrats and progressives within and without Congress repeatedly reminded Obama to remember the promises he made on Iraq and Afghanistan. In the case of Iraq, he blasted it as a failed and flawed war that should never have been fought. He promised that as president he would move as quickly as possible to end it. He kept that promise.

As for Afghanistan, he escalated the war, but there was always the explicit understanding that the established timetable for phased withdrawal would be kept and there would be an actual end to direct U.S. military involvement in the country. This still seemed like waffling on getting the U.S. out of Afghanistan.

Then there was the low intensity drone war that Obama waged against terrorist havens in Somalia, Yemen, Pakistan, and a few other Middle East hot spots. This tactic drove progressives up the wall. Things came to a head in 2013 when thirteen faith based and liberal advocacy groups sent a letter to Obama imploreding him to end the drone program. They insisted that the program blatantly violated international law, killed hundreds of innocent civilians, whose lives were more than the sanitized and euphemistic term, "collateral damage," that the military incessantly claimed.

Equally important, turning the drone program into a stealth killing machine cast Obama in the role of military hardliner, which seemingly contradicted, if not betrayed, every progressive precept and principle that he espoused. The program was indeed secretive, largely lacking any accountability, and had killed several thousand persons in these countries. How many of the dead posed a genuine terrorist threat to the U.S. would always be subject to severe questioning of. Obama would be repeatedly questioned by many about this, and deservingly so.

For his part, he said little about the program, other than to make an offer to compensate some of the families of civilians who were killed in the attacks and not deemed terrorists. The program would remain shrouded in secrecy during his White House tenure. The public would remain in the dark about the human cost and the legality of raining death on civilians from the air. For progressives, this would always be another of the blights that stained Obama's record and in some ways, even his legacy.

Though he took much heat from the left for his willingness to play the tough guy on defense and national security issues, the truth was that he moved with far more caution and veered more toward disengagement rather than confrontation on these issues than critics claimed. The Syrian strike threat in 2015 was the best example. Obama made it clear there would not be direct U.S. military involvement. This was an easy call since few Americans would have backed such a move. He hedged on when the threatened missile strikes against Syria would occur, when he said that there was no set timetable for the launch if Congress had approved action. This was his nod to his backers who opposed any action against the country, or who demanded the most limited action possible to avoid a repeat of anther Iraq and Afghanistan quagmire. Obama's challenge was to assure them a Syria strike wouldn't lead to that.

* * * * *

The health care retreat and other Obama shifts weren't the things that were making droves of progressives shut down on him throughout his presidency. Their great fear was that he'd perform more infuriating back flips or do little to expand civil rights and liberties protections, gay rights, end the wars, and rein in corporations and Wall Street abuses. They further feared that he'd make compromise and caution the twin watchwords of his administration.

There was little he could do to convince Nader and countless others that he could ever be a true progressive. The art and reality of White House governance, and the endless compromises that that required, made that all but impossible.

This was not a case of Obama being a con man, as Nader called him, but doing what presidents had to do and that was to appease, placate, and court the entrenched and powerful political opponents and interests in order to have any hope of getting anything done.

The vehement perception of Obama as betrayer of the progressive cause didn't hurt him. His popularity numbers moved back up during the last year of his presidency. It did, though, have devastating consequences for the candidate who sought to succeed him in the Oval Office and vowed to uphold the Obama legacy. That was Hillary Clinton in 2016.

There was almost a compact among progressives from the moment that Clinton announced her candidacy in April, 2015, that they would not make the same "mistake" in supporting her as they had made with Obama.

The "anybody-but-Clinton" refrain began with a long and loud call for the progressive icon and Massachusetts Senator, Elizabeth Warren, to run for president. When she declined to run, the chant became even louder for Bernie

Sanders to run. He did and millions rallied to his radical banner. He was the great radical hope that they lusted for. When Sanders didn't get the Democratic nomination, untold numbers of progressives didn't hesitate to say there would not be another Obama moment with Clinton.

Prominent actress Susan Sarandon and other progressives vowed not to back her under any circumstances. They refused to heed exhortations that staying at home on Election Day and not backing Clinton could put Trump in the White House. This would put all the gains of the past half century in women's rights, civil rights, and environmental issues at mortal risk. That a Trump win could mean a hard-right takeover of the Supreme Court, unfettered power to corporations and Wall Street, and the heightened threat of war, also had no traction with the anti- Clinton and anti-Obama hardliners on the left.

The nightmarish scenario of a Trump White House victory emerged in hellish reality. How much the progressive's pox on both Clinton and Trump's houses stance aided and abetted his win would remain the subject of fierce debate for years to come. The certainty, though, was that progressives in saying "not another Obama with Clinton," still saw Obama as the gold standard for compromise and caution and wanted no part of that again. It was a dubious reading of what Obama was able to accomplish in office, but it was their reading nonetheless.

CHAPTER 12

Was a Black President Black Enough?

From the moment that Obama set foot in the White House in January, 2009, he had to hear the charge that he miserably failed blacks. He repeatedly heard it for two reasons. The first was obvious. He's black. But more than that, he practiced civil rights law and worked as a community organizer in poor communities on Chicago's South Side. Therefore, he was expected to have an even greater sensitivity to the plight of the black poor. And because of that he had a duty, obligation, and, even a mission, to use the power and prestige of his office to do and say more for blacks.

The other reason was pragmatic. Blacks backed him in record numbers in his 2008 and 2012 election campaigns. Despite the relentless beating he took from the GOP, the legions of bloggers, websites, talk show jocks, tea party leaders and followers, and assorted racist and fringe hate groups for six years, blacks remained his staunchest supporters even when they privately grumbled that he should do more for them, especially during his second term, since he no longer had to seek reelection. Obama did, on occasion, remind the black critics that he was not the Black President but the American President. This declaration always got much ink, and ignited public and private criticism that he was not doing enough. He had some answers to the avalanche of criticism.

One was intensely personal. Following the acquittal of George Zimmerman in 2013, who was charged with gunning down black teen Trayvon Martin in Florida, Obama movingly spoke of the "pain" most African-Americans felt at the acquittal of Zimmerman. He didn't speak solely because he felt obliged to make a generic observation about the anger of most blacks toward the verdict, or even out of remembrance of the fight he led in the Illinois state legislature more than a decade before the Zimmerman acquittal to get a bill passed that put law enforcement on notice that racial profiling would not be tolerated. It took many tries and four years to get the bill finally passed.

He also spoke from a well-documented personal experience. He bluntly noted in his 2006 book *The Audacity of Hope* that he was not always a high-profile, respected and acclaimed public official. He recalled that there was a time in the not so distant past that he suffered, as he put it, "the litany" of slights and abuses that ranged the gamut. He ticked them off quickly: security guards tailing him as he shopped in department stores, white couples who tossed him their car keys as he stood outside a restaurant waiting for the valet, police cars pulling him over for no apparent reason.

Race for him was much more than a series of unpleasant, personal remembrances about personal slights and perils. He reminded his critics that he proposed, backed, implemented, enacted, and increased funding for initiatives and legislation and signed executive orders that boosted blacks in education and business. The number of low-income blacks that were eventually enrolled in the insurance exchanges with subsidies, more than bore out that his signature Affordable Care Act gave them first time access to affordable health care. Being insured gave thousands of blacks who faced life threatening illnesses a new lease of life. Obamacare, and his other initiatives that benefited blacks, mostly flew low under the radar scope, were ignored, deliberately minimized, or distorted by Obama's black critics. The reason for that was simple. Obama steadfastly refused to say that any of his initiatives and programs were aimed at blacks. The absence of this frank statement and the longing of many blacks to hear him make that pronouncement fueled the perception and frustration that he wasn't doing enough, or even anything.

Meanwhile, the GOP's black attack surrogates delighted in lashing Obama with alleged figures that showed that blacks supposedly fell further down the economic ladder during his administration than under Bush. This was crass and a cynical cherry picking of statistics to make a trumped-up case that Obama was no friend of blacks, and, therefore, they should be no friend of his. The facts belied their charge.

Studies that measured how blacks fared under GOP versus Democratic presidential administrations, including the first two years of Obama's tenure showed that blacks did increase their annual average income, including black family income, and employment and that their poverty rate decreased under Democratic presidents.

* * * * *

The gains blacks made under Obama were even more remarkable, given the other historic cross that he had to bear. For a quarter century before Bill Clinton won the White House in 1992, Democrats were regarded and reviled by conservatives as the party that tilted to and pandered to minorities. The backlash was swift and devastating. Blue collar and rural white males deserted the Democrats in droves. Their sprint to the GOP became the reliable ace card for Reagan, Bush Sr., and George W. Bush's White House wins.

Yet, Obama made creating jobs and bolstering health care, priorities even while having to maintain his presidency as race-neutral. This approach angered some blacks, but it did not give the Obama baiters anywhere to go with the issue of race. The end game was not what appearance Obama had to show to the public, but what tangible gains blacks made. Despite the criticism, Obama could rightly claim that much was done on this score even when he heard, as he would always hear, during his White House years, the charge that he failed blacks.

This was hardly the only area of complaint he had to

hear of his supposed failure to say and do more on race. The single biggest gauge that blacks used to measure how much a black president was willing to do on a racial issue was how he dealt with the issues of mass incarceration and police violence. It took Obama some time and much care, knowing just how volatile the issue of crime and punishment, law and order, was to conservatives, to nudge either issue toward the front of his public agenda.

The issue he picked to stake out turf in this policy arena was the insane, human and financial drain of the nation's draconian drug sentencing laws. He eased into it by gradually increasing the number of clemencies that he granted to mostly low-level drug offenders, and then accelerating the rate each go-round of clemencies. The prospect of such prisoners being released without the President's intervention would have been slim to none. The scorecard for the Obama administration on clemency for low level offenders, for the most part, related to drugs. By the end of his White House tenure, the number freed totaled in the hundreds.

He further drove the point home about the over incarceration of petty offenders with his presidential groundbreaking visit to a federal penitentiary in Oklahoma in July, 2015. Obama's actions were heart-lifting news for the offenders and their families, and it was great news for a nation that had acquiesced far too long in the thoroughly debunked notion that the nation could incarcerate its way out of the drug morass. Yet, as Obama certainly knew there was still much to be done for the nation to dig out of this morass. The incessant demand would remain for him to do even more.

His clemencies for drug crimes followed hard on the heels of Attorney General Eric Holder's virtual demand that U.S. Attorneys in August, 2013, rethink how and who they prosecuted for drug crimes. That followed even closer on the heels of Congress's passage of the Fair Sentencing Act in 2010. Before that, the Supreme Court issued a ruling that modified the draconian sentence for a convicted cocaine peddler. Their actions wiped out much of the disparity in the blatantly racially tinged sentences slapped on crack cocaine users.

The drug sentencing disparities had become a national embarrassment. But they were still on the books. They had wreaked dire havoc in mostly poor black communities, as well as cast an ugly glare on the failed and flawed war on drugs. Countless studies showed that blacks made up the overwhelming majority of those sentenced in federal court for crack cocaine use and sale. Contrary to popular myth and drug warrior propaganda, more than half of crack users were white, and presumably a good portion of them were crack dealers as well. It was the heart-wrenching tales of the legions of poor young men and women, though, that received sentences totaling decades behind bars for the possession or sale of a pittance of cocaine or marijuana that touched a raw nerve. In many cases, they were young mothers and fathers, who, out of poverty and desperation, resorted to the use and sale of drugs.

The result of that bloated, grotesque drug war was that

the U.S., with five percent of the planet's population, had nearly 25 percent of its inmates. The even more brutal fact that Obama could, and should, not have ignored was that the overwhelming majority of those jailed were black.

* * * * *

The one racial issue, however, that made drugs and jobs almost pale in comparison and riveted the eyes of a nation on the White House was the spate of police killings of mostly unarmed blacks in 2015 and 2016. This was capped by the back to back police slayings of Alton Sterling in Baton Rouge police and Philando Castile in Falcon Heights, Minnesota in July, 2016. The spiral of violence triggered the equally horrific slaying of five Dallas police officers the same month. At the time, Obama was on a state visit to Spain to attend an interfaith service. He cut that short to fly to Dallas to give aid and comfort to their families. This was the right thing to do. The slaying of the officers was, just as Obama called it, vicious and despicable. They were public servants, guardians of the law, and the killing was a brutal and frontal challenge to state authority. No president could ignore that threat. However, the crisis and challenge was the perfect opportunity to tackle the endless issue of race and whether black lives are less valued than those of others. Obama did not ignore the horrendous killings of Sterling and Castile. As he did in other slayings of young blacks, such as Trayvon Martin and Michael Brown in Ferguson, Missouri, he expressed agony over their deaths and the polarization and insensitivity to the lives of blacks. Yet this was not the same as personally visiting the scene of the tragedy to address this issue. Obama established a precedent for personally flying to memorials and tributes and giving aid and comfort to the victims and families of those slain in mass shootings in Aurora, Colorado, Newtown, Connecticut; San Bernardino, California; and Orlando, Florida.

He was applauded for these visits. It was the right thing to do. However, police violence against unarmed young blacks was a relentless and polarizing plague. He pled, cajoled, and prodded the nation and even law enforcement to take proactive, reform steps to curb violence by getting the Justice Department more directly involved. The Justice Department responded by holding endless meetings, conferences with local law enforcement agencies, and issuing guidelines and proposals that covered the gamut from better police training, stepped up minority recruitment, and eliminating racial profiling. Police violence in Baton Rouge and Falcon Heights and dozens of other cities was not a garden variety local issue. It was a national crisis for the very reason that the perpetrators acted under the authority of the state and law. This was not a lose-lose political minefield for Obama to step into. Their deaths were crucial public policy concern and there was a need to provide redress for a situation that had gone badly awry. It was true that presidential statements on a controversial issue have, at times, polarized and fueled political backlash. In fact, the Sterling and Castile slayings were a near textbook example of the fury and passion that occur when race is involved. There was rarely a moment during his tenure in the White House that he wasn't reminded of that. Early in his administration, he lightly ventured into a racially charged local issue with his mild rebuke of the white officer that cuffed Harvard professor Henry Louis Gates in 2009. The reaction was instant and rabid. Polls after his mild rebuke showed that many whites condemned him for backing Gates and, even more ominously, expressed big doubts about his policies. The police murders in Falcon Heights and Baton Rouge had profound racial implications. They raised painful questions. Why did so many police officers wantonly slay young blacks? Why did police and public officials often seem helpless to stop the killings? Why were black males so relentlessly typecast as violent prone criminals that made them even greater sitting ducks for police violence? Why was it in the rarest of rare case where prosecutors bring charges against police officers that wantonly kill unarmed blacks? The Sterling and Castile murders cast another horrific glare on the gaping racial disparities in how the lives of young blacks were publicly viewed and treated. They burst open even wider the perilous divisions, polarization, and seeds of violence that the Dallas police slayings horrifically showed exist. No president could afford to ignore that peril. Obama, partially moved by those tragic events and by his personal experiences as a black man and the need to translate those experiences into public policy were powerful motivating forces to compel a president to act. Obama did. Could he have done more on police violence, drugs and mass incarceration, and the crisis of black unemployment?

The case could always be made that he could. He could have authorized more aggressive Justice Department civil rights prosecutions of police officers who killed unarmed blacks when the state refused to prosecute them. He could have pressed for a full halt to the federal government's sale of military hardware to local police departments, and launched a stronger administration push for more congressional action on fair sentencing laws. He could have used the presidency as a bully pulpit to advocate the total erasure of the gaping racial disparities and prosecutions for drug offenses. He could have proposed an administration sponsored crash program on spending specifically for job and skills training that targeted young black males.

On the other hand, an equally strong case could be made that he made more than a token effort in these crucial areas. He did speak out when he could and cajoled a Congress to the extent possible to act. That is what a president should do to show that race still does matter in America. In this respect, Obama made a good first start here, a start that future administrations would hopefully build on.

CHAPTER 13

In the Racial Cross Hairs

One year into Obama's second term, and in the space of one week, in August, 2013, a protester waved a sign, "bye bye black sheep," and a small chorus chimed in and put it to the popular song ditty of "Bye, Bye, Blackbird" in front of Desert Vista High School in Phoenix where he had just spoken there about housing finance reform. Hundreds of attendees at a Missouri state fair roared with laughter and applause at a rodeo clown's mocking Obama. In Orlando, a knot of protesters waved racially insulting signs including, "Kenyan Go Home" at Obama's motorcade. And a GOP congressman hinted that there was sentiment among House Republicans for an Obama impeachment drive over questions, yet again, about his American citizenship.

These despicable and outrageous displays of racism should have been condemned. At least, though, they were honest in that the upfront bigots had shredded the carefully honed script that had been the template for the racially sneaky and insidious attacks on Obama. The script read like this: Whenever an elected official, Tea Party figure, or even leader, or gobs of protesters waved their racially loaded signs, posters, and shouted racially derogatory barbs at Obama, Tea Party and GOP leaders chalked it up to variously a fringe few, or cranks, and then wailed that it was blatantly unfair to, as well as politically conniving, to tar the Tea Party, let alone, the GOP as racist for the acts of a kooky few. They had a legitimate gripe except for one, actually, two small points.

In 2012 a Brown University social scientist used polling experiments and an independent survey to identify a series of issues that were bitter and contentious between Obama and the GOP and that appeared to be race neutral such as tax policy, health care reform, Supreme Court appointments, and political party identification. He found that those with a racial antipathy toward blacks were more prone to oppose anything that Obama supported.

There was method to the madness there. The relentless racial assault on Obama was the one unyielding constant from the moment that he declared his presidential candidacy in February, 2007. This was not simply an unorganized, spontaneous expression of bigotry. It was subtly stoked and orchestrated by the GOP with the clear political aim of disrupting, destabilizing, and rendering politically impotent Obama's programs, initiatives and proposed legislation.

* * * * *

The final presidential vote in 2008 gave plenty of warning of the lethalness of the GOP's core conservative white constituency when aroused. Overall, Obama garnered slightly more than 40 percent of the white male vote. Among Southern and Heartland America white male voters, Obama made almost no impact. The only thing that even made Obama's showing respectable in those states was the record turnout and percentage of black votes that he got. They were all Democratic votes.

A Harvard post-election assessment of the 2008 presidential vote found that race did factor into the presidential election and that it cost Obama an added three to five percent of the national popular vote. Put bluntly, if Obama had been white, the election would have been a rout. During the GOP presidential primary campaign, GOP presidential candidates made sure of that with the stream of race-tinged references Newt Gingrich, Rick Santorum, and Mitt Romney made to food stamps, welfare, work ethics, and an entitlement society. Then there were the racially-loaded newsletters from Ron Paul that resurfaced. The candidates when challenged ducked, dodged, and denied any racial intent, or in the case of Paul's newsletter, that he even penned them.

His 2012 reelection victory gave even more warning that little had changed. In fact, it got worse. Obama garnered a smaller percentage of the overall white vote than he did in 2008, and that included a small but significant defection of younger white voters who backed him in 2008. The Tea Party took much heat and continued to take some for race-baiting. The attacks stung. It retooled, repackaged, and refined its attacks and slogans when Obama and Congress locked horns over the deficit, the budget, spending cuts, and alleged scandals that plagued Attorney General Holder. These tactics were carefully designed to yank the albatross of race off the Tea Party and the GOP's back.

That hardly meant that race was off the GOP table as a major political weapon in the GOP's political attack arsenal. They had it both ways. Their shock troops in the streets in front of the high school where Obama spoke in Phoenix hurled their racist taunts at Obama, and the GOP and Tea Party leaders at the same time either stayed mute, or engaged in plausible deniability — namely: that was "just a few lone nuts, not us, that were racist."

The serenading of Obama to the tune of "Bye, Bye, Blackbird", and other clownish racist antics, simply reconfirmed the horrid fact that Obama's public policy stances and battles would continue to be dogged by subtle and overt racial attacks from the congenital Obama race baiters. Despite the fevered denials that race had anything to do with their Obama animus, there was political method to the madness. This was the burden that came with the turf for an administration in which race was never far from the surface. This was an added testament to the thin line that no other president before Obama had to walk during his every moment in the Oval Office.

* * * * *

The racism was only one side of the line that he had to walk. On the other side, there was the giant shadow of Dr. Martin Luther King, Jr. that hung over his administration. Invariably his name was linked with King. So, it was fitting that on the 50th anniversary of the March on Washington in August, 2013, Obama punctuated the official White House proclamation with the oft-quoted line from Martin Luther King Jr.'s, "I Have a Dream" speech imploring the nation to "let justice roll down like a mighty stream."

Obama did not invoke the even more oft-quoted King admonition from the speech that the content of character should negate skin color. That was telling. The first line was a frontal challenge to the nation to wage tireless battle against racial discrimination and the equally corrosive injustices of racial violence, poverty, and war. King's fight was against all three which eventually made him a pariah at the White House and the target of the FBI's decades long ruthless, relentless no-holds barred assault on him.

Fifty years after the March, Obama was the target of the same relentless, ruthless and no holds barred assault from the GOP, coupled with an even more racially skewed, snide and vengeful assault by right wing blogs, websites, and talk show hosts. Obama was the recipient of more death threats than any other modern day president. In that sense, he was right to cite King's admonishment that the job the civil rights leaders started was far from over.

That was only part of the five-decade saga of racial struggle and change that Obama benefited from and at times was throttled by. His election did show that millions of whites were more than willing to punch the ticket for an African-American for the world's most powerful political post. King would almost certainly have glowed with approval at that. Yet, there were a couple of troubling caveats that marred America's great racial leap forward.

Obama won in large part because he did what no other Democratic presidential candidate did, including Bill Clinton. He turned his presidential campaign into a virtual holy crusade by African-Americans voters to get him in the White House. At the same time, the majority of whites in 2008, and even more so in 2012, voted for his GOP rivals. They weren't all Republicans; many were rural and blue collar whites, many of whom were Democrats. Both presidential elections showed that the racial fault lines were still tightly drawn within a wide segment of the electorate.

Periodic surveys of racial attitudes during his White House years found that a significant percentage of whites who said they had voted for Obama also said that blacks were more crime prone and less industrious than whites. Other surveys including the Pew Center survey on race relations that was timed for release to coincide with the fiftieth anniversary commemoration of the March found the same racial disparities in income, wealth, housing, employment and the criminal justice system that plagued the nation in 1963. Countless government reports and studies, and the National Urban League's annual State of Black America reports also found that discrimination and poverty were still major barriers for millions. King's special concern was poverty and how it had divided the country into two Americas, one wealthy, prosperous, and in total control of the nation's wealth and resources and the other America, poor, desperate and disproportionately minority. President Obama took much heat from black critics for not forcefully addressing the needs of that America, and more importantly, as the critics claimed, not taking stronger, direct steps to alleviate the suffering. King almost certainly would have been careful not to lambaste Obama for the endemic poverty of millions or suggest that the president's policies had anything to do with that poverty. This would have just played into the hands of the arch enemies of Obama, while exacerbating their animosity toward greater civil rights protections and ending income inequality.

If King had lived, almost certainly he would have been in the streets marching for economic justice, even during Obama's tenure. His willingness to protest this in 1968, not only made him *persona non grata* at the White House, it also put much distance between him and the established civil rights organizations of that day. That almost certainly would not have been the case if King had protested wealth inequality and poverty with Obama in the White House. He would not have been excluded from access to the president. Obama frequently met with civil rights leader, Al Sharpton. He also had White House meetings with many other civil rights leaders and activists including Black Lives Matter.

Obama bristled at the notion that the civil rights movement was outdated, or worse, that he somehow supplanted the ongoing work of civil rights leaders. He repeatedly praised civil rights leaders for their heroic battle against racial injustice. He did it in his official March commemoration proclamation and at other times during his time in office. By not making any mention of King's dream of a color free and economic justice assured nation, Obama conceded the obvious; that he and America were still seeking to fulfill the dream King had a half century ago.

Every president after King paid some lip service to fulfilling King's dream, including conservative GOP presidents Reagan and Bush. What made Obama unique was that he did not just pay tribute to King's dream and legacy, but acknowledged that the dream was still unfilled, and that his historic election as the nation's first black president, was profound, yet no substitute for ending the towering racial inequities that King fought to end and that still plagued the nation during his years in the White House. The issue of race would continue to plague the nation after he left it.

CHAPTER 14

What to Do About the Poor

It was a near surreal moment in January, 2015, when two men who weren't running for President in 2016 took swipes at each other over the one issue that has been one of the major sore points in American politics; that is, what to do about poverty in America. Obama's 2012 GOP presidential opponent, Mitt Romney, touched off the sniping over poverty in a speech at Mississippi State University in which he slammed Obama for not doing enough, as he put it, to "lift people out of poverty." Obama quickly hit back in a talk to the House Democratic retreat, ridiculing Romney for "suddenly being deeply concerned about poverty." Obama then laid down a gauntlet of sorts when he challenged, "Let's do something about it."

Just exactly what that "something" was Obama didn't say. And Romney's prescription for "lifting" the poor out of their poverty by "restoring opportunity, particularly for the middle-class," had absolutely nothing to do with the poor. However, the fact that both Romney and Obama sparred over the issue of poverty and what could be done about it was telling.

However, the spat between them, for that moment, anyway, did cast a glare on what Obama had done to tackle the always thorny issue of what to do about America's poor. That question was further driven home in July, 2012, when two reports were issued the same day that told of two appalling realities. One report on poverty found that the poor were indeed getting poorer and more numerous than at any time in the last half century.

The other report from the Tax Justice Network found that the super-rich were getting richer, and they were squirreling their money away to the tune of trillions in offshore tax havens that were far outside the reach of the U.S. and other nation's tax collectors. Wealthy Americans were amply represented among the offshore tax evaders. These clandestine funds could have bankrolled business start-ups, business expansion, fattened federal and state tax revenues, and created thousands of new jobs. This would have done much to blunt the steady march upwards of those who slipped into the poverty ranks.

The devastating impact of poverty on American economic life has long been well known. It wastes the talents, energy, and productive potential of many in the work force. In some communities, it increases crime which overburdens the police, the courts, and prisons, and makes doing business in these areas costlier. It strains the health care, and the welfare system. This results in a bigger tax drain on the middle-class. It sharply reduces the ability of thousands of consumers to purchase goods and services. This further crimps business growth and reduces government tax revenues.

There was a brief glimmer of hope during the 2008 presidential campaign that then Democratic presidential contender Obama would put the fight against poverty on the front burner of his agenda. He and the other Democratic contenders, Hillary Clinton and John Edwards, thundered in speeches in April 2007 that America had failed the poor, and, especially, the black poor. With the exception of Edwards, whose candidacy soon disintegrated after public revelations about his love tryst, Obama and Clinton didn't utter another word about poverty for the rest of the campaign.

The 2012 presidential campaign was no different. There was not even a faint mention of the word poverty on the presidential campaign trail by either Romney or Obama. There was a reason; in fact several reasons why. One was trying to define who is poor. Apart from the visibly homeless, and those rummaging around on skid row, and in some of the poorest and most recognizable urban inner city communities, one could easily be considered working, or even middle class, one day. Then loss of a job and tangible income could quickly dump that person into the poverty ranks the next day. This made the poor even more diffuse and harder to typecast. They cut across all ethnic, gender, and religious, and even political party affiliation lines. There were low income persons in the South, Middle America, and the rural areas, who were conservative, and voted GOP.

* * * * *

The other reason was that the poor do not have an advocacy group to go to bat for them with lawmakers such as labor, civil rights, education, environmental, or abortion rights supporters have. This further increased their political invisibility. The only time the poor had loud champions was a moment during the 1960s when a small band of anti-poverty groups and organizers got the attention of the Johnson Administration. They shouted, cajoled, and actively lobbied LBJ for a major expansion of anti-poverty programs, funding, and initiatives to reduce poverty in the nation. But the anti-poverty crusade quickly fell victim to Johnson's Vietnam War build up, and the increasingly shrill attacks from conservatives that the war on poverty was a scam to reward deadbeats and loafers, which triggered sharp budget cutbacks.

Another reason for the silence about poverty on the campaign trail was that the national economic and fiscal emphasis has been on how to hack away tens of billions from spending on domestic and even, the once sacred, military programs. Obama and Romney did fiercely arm wrestle during the 2012 presidential campaign over which one could best bring down the deficit, reduce spending, decrease taxes, and get rid of wasteful programs. They also sparred hard over how best to protect the interests of the middle-class. It was about votes, and a pro-middle-class, cut spending line, appealed to centrist, and independents. Both Romney and Obama banked on the two groups to be the path to, or return to, the White House.

The biggest reason that Obama did not dare to make poverty a political issue in 2012 was the same reason that no other presidential contender has done so since LBJ. The existence of so many poor people flies in the face of the embedded laissez faire notion that the poor aren't poor because of the hyper concentration of wealth, or worse, any failing of the system, but because of their personal failings. Surveys have borne this out. Even many among the impoverished were as apt as many of those in the middle-class, and the well-to-do, to self-debase themselves for their poverty. They blamed it on their misfortune, bad luck, lack of education and skills, or alcohol and drug problems. These were certainly reasons why some fall into poverty or remain chronically poor. These reasons, however, were at best, peripheral to the real cause of the increase in poverty. That's the control by a relatively handful of the bulk of the nation's income, resources and productive wealth.

* * * * *

No matter the how the public and policy makers perceived the causes of poverty, Obama was still wide open to the charge that the poor were growing by leaps and bounds and somehow he should be blamed for this. The reality for Obama on this was two-fold. He was the man at the top and fairly or unfairly, he would have to take the heat for the poverty rise, particularly since the GOP had done everything to tie the issue into his economic policies, which they relentlessly pounded as ineffectual. Obama took the cue. His economic stimulus plan in 2009 didn't earmark any dollars specifically for anti-poverty programs or projects. There was no indication that Obama or Congress would have proposed any new initiatives, or increased funding specifically targeting chronic minority unemployment, or for low income family support programs.

However, the other reality was that those same policies, particularly his 2009 spending stimulus program that ladled out billions and provided for larger tax credits, emergency unemployment benefits, and expanded food assistance, a program, that the GOP ripped, did broaden the safety net. The Center for Budget and Policy Priorities estimated that the Recovery Act ultimately kept about seven million people above the poverty line. This was the life saver for many that Obama could take credit for.

The dangling question, though, was could he have done more to alleviate the hardship of the poor? The answer was yes and no. He could have pushed for an even bigger stimulus spending program. There were many liberal Democrats and progressives who screamed at the president to do just that, and when he didn't, they hammered him. The cold fact was that the GOP controlled House would have rendered a bigger spending plan DOA. It was a classic case of a president who would have wanted to do more bumping up against a Congress that out of pure political animus wanted to do even less for the poor.

The big loser was the millions that continued to wallow at the bottom of America's economic ladder. They would remain victims of America's structural economic inequality, the dislike, denial of, and blind eye by many Americans to the poor, and politician's traditional disdain for them. Obama didn't create this tormenting and seemingly intractable confluence of terrible forces. Yet, those forces ultimately tied his hands on just how much he could have ever done for the millions that would remain poor despite his best efforts.

CHAPTER 15

Surviving the GOP's One Term Presidency Vow

GOP Senate majority leader Mitch McConnell drew much derision, snickers, and ridicule following Obama's re-election in 2012 when he fell flat in fulfilling his oft repeated goal to make Obama a one-term president. Obama's smash electoral vote and solid popular vote win made the GOP's aim seem more comical. The fiscal cliff battle over the budget in 2009 and 2010 seemed to render it even more absurd. Obama was the big winner. He preserved the Bush tax cuts for the middle and working class, got a big portion of cuts scrapped for the highest income earners, took Social Security, Medicare and Medicaid off the budget slice table, delayed debate on proposed draconian GOP budget cuts that would have slammed the poor and needy, and delayed the battle over the debt ceiling.

Yet that was just the opening gun of what the GOP loudly

claimed then would be a bitter, protracted and divisive fight to stymie Obama's second term presidency. The GOP was under even more intense pressure from ultra conservatives who had screamed at those GOP congresspersons for supposedly giving too much away to Obama in the fiscal cliff debate and not taking the hardest line stance they could against deals with his White House. GOP House leaders, reeling from the criticism, said they wouldn't even consider talking with Obama behind closed doors in the future to try to iron out their differences. McConnell went further and couched his war like challenge to Obama as a battle to save the nation's future.

The GOP had several formidable weapons to hammer Obama. One was its power to say no in Congress. It did that repeatedly in the Senate on one issue that flew under the public's radar scope. It refused to confirm legions of Obama's judicial appointees. He had fewer of his judicial picks confirmed then any first term president in the last quarter century, including even district court nominees. In years past their confirmation had been routine. The GOP gave no hint that it would reverse course and approve many of his picks in the future. Conservative judges showed that they could obstruct or torpedo Obama administration policies with their anti-environmental, anti-gun control, and pro-corporate rulings. GOP leaders also showed that they would harass, harangue, and try to intimidate Obama on his cabinet picks.

The GOP's assault on the possible nomination of UN ambassador Susan Rice to replace Hillary Clinton as Secretary of State in 2012 was a prime example of their intransigence. Rice's withdrawal from consideration, under pressure, emboldened the GOP to use the tactic to constantly send the message that it had the political muscle to get some of its way on public policy issues, and to keep the Obama administration off balance. Its next target that year was Chuck Hagel, Obama's choice for Defense Secretary. Despite much blustering and posturing about Hagel's credentials for the post from some GOP Senators, he was a solid establishment pick. He was quickly confirmed.

* * * * *

The GOP also could dither, delay, and obstruct implementation and funding of health care reforms, tax and budget proposals, and regulatory reforms that still needed bipartisan cooperation to pass each budget year cycle. This was crucial since Obama needed to strike deals and make compromises with the GOP to get anything done in Congress. That necessity was even more compelling, given the constant, rancorous battle over the debt ceiling, immigration reform, spending cuts, and reforms of Social Security, Medicare and Medicaid that was almost ritual for several years between the GOP controlled House and Obama. A prolonged war with the GOP that resulted in the White House getting little or none of its legislative agenda through Congress always ran the risk of souring public opinion, not just on the GOP, but the White House. This was always the bane of presidents during their second term, and, at times, marred their legacy.

There was also the power of money. Many well-heeled corporate bankrolled candidates went down to defeat in the 2012 presidential election. Some saw this as a grand rejection of the corporate, banking, and wealthy ultra conservative bankrollers ability to buy their way into office with their handpicked conservative candidates. That was misreading. Money would continue to be a potent weapon at the GOP's disposal. The astronomical cost of winning an office virtually assured that. The difference would be that in future elections, the GOP would be more selective and prudent about the conservatives whom its campaign financiers bankrolled.

The GOP's greatest weapon was the frozen political divide in the country. Nearly fifty percent of the nation's voters not only did not support Obama in 2012, but expressed total contempt for his policies and his administration. The GOP banked that it could swivel that divisiveness into sustained opposition to those policies, and that it could buy enough time with that until the 2014 midterm elections and further boost its numbers in the House, especially the Senate. As it turned out, it was a good gamble, the GOP increased its stranglehold on the House and took back control of the Senate in 2014, guaranteeing that the war against Obama's initiatives would both harden and escalate.

The 2012 election defeat did not in any way sate the GOP's thirst to regain its dominance in national politics, including its ultimate goal of winning back the White House. If it could succeed in tarring Obama as a failed president, then it cynically calculated that that would make the party's goal much easier to attain.

* * * * *

If anyone had said in 2008 after his historic White House win, that Obama's popularity rating would nosedive to the dreadful level of George W. Bush's ratings during the last years of his presidency, they would have been fitted for a straight-jacket. Obama's popularity ratings at that point had soared past 70 percent, and there was the firm consensus that his numbers would stay comfortably high. No matter how rocky things got during his tenure, they could never bottom out to Bush's abysmal numbers.

Unfortunately, some polls did show that the worst happened, and that Obama's popularity rating during the middle of his second term were virtually identical with Bush's low rating. The added insult was that Bush seemed to be getting more popular with his numbers on the uptick. There were two ways to account for that. One was that Bush had sunk so low in popularity ratings by the time he left office that he had nowhere to go but up and that it was easy for the public to wax nostalgically about and to even find a few good things to say about an ex-president years removed from office as opposed to a president who sits in the office. This view seemed more plausible given the constant barrage from the GOP's inveterate Obama bashers. They played up Bush's alleged accomplishments while relentlessly pile driving Obama's supposed failures.

The GOP's backdoor strategy, in that regard, was to wink and nod at the dirty smear campaign from the coterie of right wing talk show hosts, bloggers and web sites, that lambasted Obama with an avalanche of subtle and outright racist digs, barbs, taunts and harangues. The dual strategy had one aim, and that was to make him a failed, one term president, and failing that, a failed presidency. The added key to making that work was to play up to the hilt any and every real or perceived stumble.

During a period in 2012 and 2013 there were some problematic situations. There was the NSA spy controversy involving illegal wire taps on American citizens, the lingering anger over Benghazi, the worry over the Affordable Health Care Act website glitches, and the militant Islamist insurgency in Iraq. They embroiled the administration in continued rancor with the GOP.

The GOP's gloat that Obama was no better than Bush fell appalling flat. Bush's miserable record on the two greatest issues that mattered the most to Americans were glaring proof of that. They were the economy and war. Bush hit the skids during his second go-round in office because of public souring on a failed, flawed, costly and human-draining war, and a financial collapse that had much to do with his disastrous two tax cuts. The tax cuts gave away the company store to corporations and the rich and sent the budget deficit skyrocketing. In contrast, Obama's fiscal and budgetary record showed steady joblessness drops, a deficit fall, and an unprecedented surge in the markets that, ironically, made more millions for many of the corporate rich who had piled onto the assault against Obama.

His wind down of the Iraq war had been a special sore point for GOP war-hawks, who never tired of telling all who'd listen, that this supposedly put Americans at horrible risk from terrorism and war. It was bunk. Obama simply fulfilled commitments that were already in place to disengage the American military in that country — commitments that were supported by the overwhelming majority of Americans.

Bush could never lay claim to these achievements. Yet, the brutal political reality for Obama was that past presidents have certainly had their share of second term woes. This was the case with Eisenhower, Nixon, Reagan, George H.W. Bush, and Clinton. This shouldn't surprise. They were in office for a relatively long time. They all ran a big sprawling government with thousands of appointees and personnel. It was simply beyond the pale of one person to control every facet and decision their appointees and personnel made. Just as time could work for a second term president, it could also work against him, too. The longer he was in office, it was almost assured that some issue, event or catastrophe would happen that could blemish a president's image.

Midway through his second term, Obama had accomplished far too much in the face of far too much adversity for that to happen. Even at the depth of his approval drop, there were clear signs that things could change. The jobless rate had inched steadily down; inflation remained low; there was no major foreign policy crisis; and no personal scandal ensnared Obama. He was also a lame duck president which made him a less inviting target. With the looming 2016 presidential election, the GOP had a far more inviting target for its withering fire, namely Hillary Clinton. The GOP's attempt to spin Obama's momentary plunging popularity numbers as if he was a complete failure was politically cynical and crass.

The Obama Legacy

On January 14, 2014, a plainly worried President Obama convened a meeting of his cabinet in the Cabinet Room of the White House. The cabinet meeting had extra special, even, grave, importance. The Democrats had lost control of the Senate which meant that Congress was now completely dominated by the GOP. Obama boldly announced to the cabinet, "We're not going to be waiting for legislation. I've got a pen and I've got a phone. And I can use that pen to sign executive orders and take executive actions and administrative actions to move the ball forward." He felt he had no choice. With the GOP in full command of Congress it had total power to do what it vowed to do from the first day in January, 2009 when Obama took office. That was to obstruct, obstruct, obstruct.

So, he was determined to do an end around the machinations of the GOP Congress and wield his power of the pen to sign executive orders that gave amnesty to some immigrants, strengthened the Clean Air Act, and health care, tightened financial regulations, imposed gun control provisions, and backed transgender bathrooms. The executive signings were in addition to the more than 600 regulations that he signed off during his eight years in office that were fully in line with his agenda and that had met fierce opposition from the GOP.

However, Obama's liberal use of the executive pen to skirt Congress raised two thorny questions and posed two perils. The first: Was would his successor in the Oval Office, Trump, make good on virtually the first promise he made after being elected in November, 2016 which was to repeal many, most, if not all, of Obama's executive orders? The other more far reaching question and peril was what impact would this have on sullying Obama's legacy of accomplishments? It was true that incoming GOP and Democratic presidents took dead aim at the executive orders their predecessors of the opposite party signed, and they scrapped some of them. But in Obama's case, Trump's threat to do away with his executive orders was even more problematic, not just because of his historic first as the nation's first black president, but because so much of Obama's presidential identity and accomplishments were tied squarely to his executive orders.

There was much at stake. Yet, this was nothing new for Obama. The instant that he tossed his hat in the presidential rink in February, 2007, the triple mantra was that he could be the first black to be president. This would be convincing proof that America had finally kicked its race syndrome. And he could cement a governing White House legacy that could be a shining model for future generations to emulate.

This triple mantra was repeated ad infinitum. It was

right and wrong about the racial implication of his election. The early hint that race was overblown and over obsessed came from Obama. He didn't talk about it. For good reason, he was not running as a black presidential aspirant. He was running as a presidential aspirant. He had to make that crucial distinction for personal and political purposes.

The ritual preface of the word "black" in front of any and every achievement or breakthrough that an African-American makes is insulting, condescending, and minimizes their achievement. It maintains and reinforces the very racial separation that much of America claims it is trying to get past. Dumping the historic burden of race on blacks measures an individual's success or failure by a group standard. That's a burden whites don't have to bear. They succeed or fail solely as individuals.

Obama's personal history—his bi-racial parents, his upbringing, his education, and his relative youth— defied racial pigeonholing. He was influenced by, but not shaped by, the rigid race grounded civil rights struggles of the 1960s as older whites and blacks were.

The institution of the presidency, and what it takes to get it, demands that racial typecasting be scrapped anyway. Obama would have had no hope of bagging the presidency if there had been the slightest hint that he embraced the race tinged politics of Al Sharpton and Jesse Jackson. His campaign would have been marginalized and compartmentalized as merely the politics of racial symbolism.

He could not have raised record amounts of campaign

cash. He would not have been fawned over by legions of Hollywood celebrities, corporate and union leaders. He would not have netted the endorsements of Colin Powell and packs of former Reagan and Bush Sr. administration stalwarts, and prepped by W. Bush political guru Karl Rove on how to beat Hillary Clinton. The media would never have given him the top heavy favorable coverage, endorsements, nor relentlessly hammered Republican rival John McCain. If the media had so chosen, it could have torpedoed Obama's campaign by playing up his connection with his race focused former pastor Jeremiah Wright. It bought his protest of racial bewilderment at the Wright race revelations and dropped the matter.

* * * * *

These pluses and minuses had a deep impact on the significance and reach of his presidency, both during his White House years and afterwards. To make it work, and be that shining model of success, he had to cling closely to the centrist blueprint former President Bill Clinton laid out for Democrats in the 1990s to win elections, and to govern after they won office.

It meant that he had to emphasize strong defense, prosecute a vigorous war against terrorism, wind down the Iraq War, push for mild tax reform for the middle-class, and a workable plan for affordable health care and advocate tougher regulations to rein in Wall Street, back comprehensive gun control legislation, jumpstart climate control measures, put the obligatory Democratic Party stress on strengthening

LGBT, women's and civil rights, and make moderate and balanced Supreme Court picks and administration posts.

He walked the narrowest of tightropes in this balancing act. Every step of the way he faced a GOP that made it clear its primary goal was to make him a one term president, and, failing that, cripple his presidency by blocking his proposed initiatives and legislation.

Obama's careful, center-governing, non-racial staff and cabinet picks and his policies were plainly designed to blunt the standard Republican rap that Democrats, especially one branded a liberal Democrat, inherently pander to special interests, i.e. minorities, are pro expansive government, and anti-business. During every moment of his eight years in the White House, the GOP watched him hawk like for any sign of that.

President Obama, as all presidents, was pulled and tugged at by corporate and defense industry lobbyists, the oil and nuclear power industry, government regulators, environmental watchdog groups, conservative family values groups, moderate and conservative GOP senators and house members, and foreign diplomats and leaders. They all had their priorities and agendas, and all vied to get White House support for their pet legislation, or to kill or cripple legislation that threatened their interests.

Yet, the Obama White House was a historic and symbolic first. It was a White House that kept a firm, cautious, and conciliatory eye on mid-America public opinion, and corporate and defense industry interests in making policy decisions and determining priorities. Obama would and could not have attained the White House if he hadn't done the same. This had nothing to do with race, or attaining the historic first of being the nation's first black president. It had everything to do with the requirement of White House governance.

Obama would be praised in the inevitable historic assessment of a president for his signature successes from the Affordable Care Act to the elimination of bin Laden and criticized on the places where his administration fell short. Those places were foremost, the absence of comprehensive gun control legislation and an unfinished war in Afghanistan. That historic assessment must be made with the full recognition that Obama, as all presidents, was constrained by the inherent limit of the power of the office, the checks imposed on him by his foes, and the lofty, but often unattainable, expectations imposed on him by friends. The Obama legacy must not be defined solely by his historic first, but rather by his successes and shortcomings. Despite the much restricted and tightly defined parameters under which presidents must operate. There is more than enough in the way of accomplishment from Obama's presidency to say that his legacy is secure.

Still, Trump's win stirred much talk and much anxiety about Obama's legacy being wrecked and that somehow Trump's win was a repudiation by Americans of his legacy. This was nonsense. Hillary Clinton got tens of thousands more popular votes than Trump. She made it clear throughout the campaign that she would keep, and even expand on, his programs if elected.

There was, though, real danger that Trump and the GOP controlled Congress could gut some, or try to gut, all of Obama's program. Obama was well-aware of this possibility. Two weeks after Trump's win, he spoke at a conference of the Asia-Pacific Economic Cooperation Summit in Peru. He made it clear that he would not be a mere sideline silent spectator if there was an attempt to dismantle his programs.

"As an American citizen who cares deeply about our country, if there are issues that have less to do with the specifics of some legislative proposals or battle but go to core questions about our values and our ideals, and if I think that it's necessary or helpful for me to defend those ideals, then I'll examine it when it comes." The clear and loud warning from this was that the one who would fight hardest to safeguard the Obama legacy, was Obama.

The 7 Speeches That Cemented the Legacy

The Speech at the Democratic National Convention July, 2004

My parents shared not only an improbable love; they shared an abiding faith in the possibilities of this nation. They would give me an African name, Barack or "blessed," believing that in a tolerant America your name is no barrier to success. They imagined me going to the best schools in the land, even though they weren't rich, because in a generous America you don't have to be rich to achieve your potential. They're both passed away now, and yet I know that on this night they look down on me with great pride. They stand here, and I stand here today, grateful for the diversity of my heritage, aware that my parents' dreams live on in my two precious daughters. I stand here knowing that my story is part of the larger American story.

I stand here today, grateful for the diversity of my heritage, aware that my parents' dreams live on in my two precious daughters. I stand here knowing that my story is part of the larger American story, that I owe a debt to all of those who came before me, and that in no other country on Earth is my story even possible.

Well, I say to them tonight, there is not a liberal America and a conservative America—there is the United States of America. There is not a black America and a white America and Latino America and Asian America—there's the United States of America.

The pundits like to slice-and-dice our country into Red States and Blue States; Red States for Republicans, Blue States for Democrats. But I've got news for them, too: We worship an awesome God in the Blue States, and we don't like federal agents poking around in our libraries in the Red States. We coach Little League in the Blue States, and, yes, we've got some gay friends in the Red States. There are patriots who opposed the war in Iraq and there are patriots who supported the war in Iraq.

It's the hope of slaves sitting around a fire singing freedom songs. The hope of immigrants setting out for distant shores. The hope of a young naval lieutenant bravely patrolling the Mekong Delta. The hope of a mill worker's son who dares to defy the odds. The hope of a skinny kid with a funny name who believes that America has a place for him, too.

Hope! Hope in the face of difficulty! Hope in the face of uncertainty! The audacity of hope! In the end, that is God's greatest gift to us, the bedrock of this nation. A belief in things not seen. A belief that there are better days ahead.

The Speech in Cairo, Egypt June, 2009

[N]one of us should tolerate these extremists. They have killed in many countries. They have killed people of different faiths—but more than any other, they have killed Muslims. Their actions are irreconcilable with the rights of human beings, the progress of nations, and with Islam. The Holy Koran teaches that whoever kills an innocent is as—it is as if he has killed all mankind. . . . Islam is not part of the problem in combating violent extremism—it is an important part of promoting peace.

Around the world, the Jewish people were persecuted for centuries, and anti-Semitism in Europe culminated in an unprecedented Holocaust. . . . Six million Jews were killed—more than the entire Jewish population of Israel today. Denying that fact is baseless, it is ignorant, and it is hateful. Threatening Israel with destruction, or repeating vile stereotypes about Jews, is deeply wrong, and only serves to evoke in the minds of Israelis this most painful of memories while preventing the peace that the people of this region deserve.

On the other hand, it is also undeniable that the Palestinian people—Muslims and Christians—have suffered in pursuit of a homeland. . . . Many wait in refugee camps in the West Bank, Gaza and neighboring lands for a life of peace and security that they have never been able to lead. They endure the daily humiliations—large and small—that come with occupation. So let there be no doubt: The situation for the Palestinian people is intolerable. . . .

For decades, then, there has been a stalemate: two peoples with legitimate aspirations, each with a painful history that makes compromise elusive. . . .The only resolution is for the aspirations of both sides to be met through two states, where Israelis and Palestinians each live in peace and security. . . .Palestinians must abandon violence. Resistance through violence and killing is wrong and it does not succeed. . . .

At the same time, Israelis must acknowledge that just as Israel's right to exist cannot be denied, neither can Palestine's. The United States does not accept the legitimacy of continued Israeli settlements. For many years, Iran has defined itself in part by its opposition to my country, and there is in fact a tumultuous history between us. . . .But it is clear to all concerned that when it comes to nuclear weapons, we have reached a decisive point. This is not simply about America's interests. It's about preventing a nuclear arms race in the Middle East that could lead this region and the world down a hugely dangerous path.

[T]here are some who advocate for democracy only when they're out of power; once in power, they are ruthless in suppressing the rights of others. So no matter where it takes hold, government of the people and by the people sets a single standard for all who would hold power: You must maintain your power through consent, not coercion; you must respect the rights of minorities, and participate with a spirit of tolerance and compromise; you must place the interests of your people and the legitimate workings of the political process above your party. Without these ingredients, elections alone do not make true democracy.

The Speech
Accepting the Nobel Peace Prize
December, 2009

We must begin by acknowledging the hard truth: We will not eradicate violent conflict in our lifetimes. There will be times when nations—acting individually or in concert—will find the use of force not only necessary but morally justified. I make this statement mindful of what Martin Luther King Jr. said in this same ceremony years ago: "Violence never brings permanent peace. It solves no social problem: it merely creates new and more complicated ones." As someone who stands here as a direct consequence of Dr. King's life work, I am living testimony to the moral force of non-violence. I know there's nothing weak—nothing passive—nothing naïve—in the creed and lives of Gandhi and King. But as a head of state sworn to protect and defend my nation, I cannot be guided by their examples alone. I face the world as it is, and cannot stand idle in the face of threats to the American people. For make no mistake: Evil does exist in the world. A non-violent movement could not have halted Hitler's armies. Negotiations cannot convince al Qaeda's leaders to lay down their arms. To say that force may sometimes be necessary is not a call to cynicism—it is a recognition of history; the imperfections of man and the limits of reason. I raise this point, I begin with this point because in many countries there is a deep ambivalence about military action today, no matter what the cause. And at times, this is joined by a reflexive suspicion of America, the world's sole military superpower. But the world must remember that it was not simply international institutions—not just treaties and declarations—that brought stability to a post-World War II world. Whatever mistakes we have made, the plain fact is this: The United States of America has helped underwrite global security for more than six decades with the blood of our citizens and the strength of our arms. The service and sacrifice of our men and women in uniform has promoted peace and prosperity from Germany to Korea, and enabled democracy to take hold in places like the Balkans. We have borne this burden not because we seek to impose our will. We have done so out of enlightened self-interest—because we seek a better future for our children and grandchildren, and we believe that their lives will be better if others' children and grandchildren can live in freedom and prosperity. So yes, the instruments of war do have a role to play in preserving the peace. And yet this truth must coexist with another—that no matter how justified, war promises human tragedy. The soldier's courage and sacrifice is full of glory, expressing devotion to country, to cause, to comrades in arms. But war itself is never glorious, and we must never trumpet it as such.

The Speech on the 50th Anniversary of the Selma March March, 2015

The march on Selma was part of a broader campaign that spanned generations; the leaders that day part of a long line of heroes. We gather here to celebrate them. We gather here to honor the courage of ordinary Americans willing to endure billy clubs and the chastening rod; tear gas and the trampling hoof; men and women who despite the gush of blood and splintered bone would stay true to their North Star and keep marching towards justice.

They did as Scripture instructed: "Rejoice in hope, be patient in tribulation, be constant in prayer." And in the days to come, they went back again and again. When the trumpet call sounded for more to join, the people came—black and white, young and old, Christian and Jew, waving the American flag and singing the same anthems full of faith and hope. A white newsman, Bill Plante, who covered the marches then and who is with us here today, quipped that the growing number of white people lowered the quality of the singing. To those who marched, though, those old gospel songs must have never sounded so sweet.

In time, their chorus would well up and reach President Johnson. And he would send them protection, and speak to the nation, echoing their call for America and the world to hear: "We shall overcome." What enormous faith these men and women had. Faith in God, but also faith in America.

The Speech at the Charleston, South Carolina Shooting Eulogy June, 2015

The church is and always has been the center of African American life... a place to call our own in a too-often hostile world, a sanctuary from so many hardships. Over the course of centuries, black churches served as hush harbors, where slaves could worship in safety, praise houses, where their free descendants could gather and shout "Hallelujah"... rest stops for the weary along the Underground Railroad, bunkers for the foot soldiers of the civil-rights movement.

They have been and continue to community centers, where we organize for jobs and justice, places of scholarship and network, places where children are loved and fed and kept out of harms way and told that they are beautiful and smart and taught that they matter.

That's what happens in church. That's what the black church means—our beating heart, the place where our dignity as a people in inviolate. There's no better example of this tradition than Mother Emanuel, a church... a church built by blacks seeking liberty, burned to the ground because its founders sought to end slavery only to rise up again, a phoenix from these ashes.

When there were laws banning all-black church gatherers, services happened here anyway in defiance of unjust laws. When there was a righteous movement to dismantle

Jim Crow, Dr. Martin Luther King, Jr. preached from its pulpit, and marches began from its steps.

A sacred place, this church, not just for blacks, not just for Christians but for every American who cares about the steady expansion… of human rights and human dignity in this country, a foundation stone for liberty and justice for all. That's what the church meant.

The Speech at Rutgers University May, 2016

America converges here. And in so many ways, the history of Rutgers mirrors the evolution of America—the course by which we became bigger, stronger, and richer and more dynamic, and a more inclusive nation. But America's progress has never been smooth or steady. Progress doesn't travel in a straight line. It zigs and zags in fits and starts. Progress in America has been hard and contentious, and sometimes bloody. It remains uneven and at times, for every two steps forward, it feels like we take one step back.

Now, for some of you, this may sound like your college career. (Laughter.) It sounds like mine, anyway. (Laughter.) Which makes sense, because measured against the whole of human history, America remains a very young nation—younger, even, than this university.

But progress is bumpy. It always has been. But because of dreamers and innovators and strivers and activists, progress has been this nation's hallmark. I'm fond of quoting Dr. Martin Luther King, Jr., who said, "The arc of the moral universe is long, but it bends towards justice." (Applause.) It bends towards justice. I believe that. But I also believe that the arc of our nation, the arc of the world does not bend towards justice, or freedom, or equality, or prosperity on its own. It depends on us, on the choices we make, particularly at certain inflection points in history; particularly when big changes are happening and everything seems up for grabs.

And, Class of 2016, you are graduating at such an inflection point. Since the start of this new millennium, you've already witnessed horrific terrorist attacks, and war, and a Great Recession. You've seen economic and technological and cultural shifts that are profoundly altering how we work and how we communicate, how we live, how we form families. The pace of change is not subsiding; it is accelerating. And these changes offer not only great opportunity, but also great peril.

The Speech at the Democratic National Convention July, 2016

A lot's happened over the years. And while this nation has been tested by war and it's been tested by recession and all manner of challenges, I stand before you again tonight, after almost two terms as your president, to tell you I am even more optimistic about the future of America than ever before. How could I not be, after all that we've achieved together? After the worst recession in 80 years, we've fought our way back. We've seen deficits come down, 401(k)s recover, an auto industry set new records, unemployment reach eight-year lows, and our businesses create 15 million new jobs.

After a century of trying, we declared that health care in America is not a privilege for a few, it is a right for everybody. After decades of talk, we finally began to wean ourselves off foreign oil, we doubled our production of clean energy. We brought more of our troops home to their families, and we delivered justice to Osama bin Laden.Through diplomacy, we shut down Iran's nuclear weapons program, we opened up a new chapter with the people of Cuba, brought nearly 200 nations together around a climate agreement that could save this planet for our children.

We put policies in place to help students with loans, protect consumers from fraud, cut veteran homelessness almost in half. And through countless acts of quiet courage, America learned that love has no limits, and marriage equality is now a reality across the land.

By so many measures, our country is stronger and more prosperous than it was when we started. And through every victory and every setback, I've insisted that change is never easy, and never quick; that we wouldn't meet all of our challenges in one term, or one presidency, or even in one lifetime.

Notes

Introduction

Erin Kelly, "Republicans seek quick repeal of latest Obama administration regulations," http://www.usatoday.com/story/news/politics/elections/2016/11/15/republicans-seek-quick-repeal-latest-obama-administration-regulations/93810344/

Kamelia Angelova, "What Occupy Wall Street Protesters Really Think Of Obama?" http://www.businessinsider.com/the-download-occupy-wall-street-obama-2011-10

Amanda Ruggeri, "Young Voters Powered Obama's Victory While Shrugging Off Slacker Image," http://www.usnews.com/news/campaign-2008/articles/2008/11/06/young-voters-powered-obamas-victory-while-shrugging-off-slacker-image

Chapter 1
Countering the Bush Failure

Matt Stoler, 'The Progressive case Against Obama," http://www.salon.com/2012/10/27/the_progressive_case_against_obama/

Paul Krugman, "Obama the Moderate Conservative," http://krugman.blogs.nytimes.com/2011/07/27/obama-the-moderate-conservative/?_r=0

News One Staff, "Democrats Want Obama to Stop Compromising with the GOP," http://newsone.com/1509175/democrats-want-obama-to-stop-compromising-with-gop/

Kimberly Amadeo, "Compare Obama vs Bush on Economic Policies and the Debt Was Bush or Obama best for the Economy?" http://www.thebalance.com/how-do-obama-and-bush-compare-on-their-economic-policies-3305622

Chapter 2
The Obamacare Presidency

"Obama eases off 'one-term proposition,' says economic fix takes more than 4 years," http://www.foxnews.com/politics/2012/11/01/obama-knew-from-beginning-it-would-take-more-than-1-term-to-fix-economy.html

Miles Mogulescu, "Obamacare's Flaws Put Progressive Politics in Jeopardy," http://www.huffingtonpost.com/miles-mogulescu/obamacares-flaws-put-prog_b_4310434.html

Dana Milbank, "The GOP's Latest Doomed Quests to Kill Obamacare," http://www.washingtonpost.com/opinions/the-gops-latest-doomed-quests-to-kill-obamacare/2015/12/02/d2502608-9947-11e5-8917-653b65c809eb_story.html?utm_term=.99e87629261c

"The Key features of the Affordable Care Act," http://www.hhs.gov/healthcare/facts-and-features/key-features-of-aca/

Patrick Temple-West, "Pence: Trump to Push Rapid Repeal of Obamacare," http://www.politico.com/story/2016/11/trump-pence-obamacare-231677

Chapter 3
Making Like FDR

NPR, "Obama and FDR: Similar Challenges," http://www.npr.org/templates/story/story.php?storyId=114376732

"Political Critiques of the New Deal are Boundless," http://www.boundless.com/u-s-history/textbooks/boundless-u-s-history-textbook/the-new-deal-1933-1940-25/critical-interpretations-of-the-new-deal-193/political-critiques-of-the-new-deal-1071-2233/

Jackie Calmes, "Obama's Budget Retains Agenda," http://www.post-gazette.com/nation/2010/02/02/Obama-budget-retains-agenda/stories/201002020219

Earl Ofari Hutchinson, "Obama Slays GOP Myth of 'Evil' Big Government," http://www.huffingtonpost.com/earl-ofari-hutchinson/president-obama-slays-gop_b_2702035.html

Chapter 4
Kicking the Foreign War Syndrome

Mark Landler, "U.S. Troops to leave Iraq by Year's End, Obama Says," http://www.nytimes.com/2011/10/22/world/middleeast/president-obama-announces-end-of-war-in-iraq.html

NY Times Editorial Board, "Iraq War Lies, 13 Years Later," http://www.nytimes.com/2016/07/08/opinion/iraq-war-lies-13-years-later.html

Justin Sink, "Obama Finds he Can't Escape the Afghan War He Once Vowed to End," http://www.bloomberg.com/politics/articles/2016-07-06/obama-scales-back-afghan-withdrawal-amid-continued-threats

"Obama's Speech at Woodrow Wilson Center," http://www.cfr.org/elections/obamas-speech-woodrow-wilson-center/p13974

PBS News Hour, "More Americans Oppose the War in Afghanistan than any other Recent American Conflict," http://www.pbs.org/newshour/rundown/more-americans-oppose-the-war-in-afghanistan-than-any-other-recent-american-conflict/

"Unpopular wars = unpopular presidents; Obama's moves on Iraq and Afghanistan as 2012 nears," http://latimesblogs.latimes.com/washington/2010/09/obama-iraq-joe-biden.html

Bradford Richardson, "Republicans Blast Obama's Syria Policy," http://thehill.com/policy/defense/258786-

GOP-criticizes-obama-for-syria-policy

"What War Powers Does the President Have," http://www.slate.com/articles/news_and_politics/explainer/2001/09/what_war_powers_does_the_president_have.html

Edward Delman, "Obama Promised to End America's Wars—Has He?" http://www.theatlantic.com/international/archive/2016/03/obama-doctrine-wars-numbers/474531/

Chapter 5
Rough Road to Benghazi

"This is what Actually Happened at the US Consulate in Benghazi," http://www.rawstory.com/2016/02/this-is-what-actually-happened-at-the-us-consulate-in-benghazi/

"No Bombshell but More Criticism of Clinton and Obama in Benghazi Report," http://www.wibc.com/news/national/no-bombshell-more-criticism-clinton-obama-benghazi-report

"Jimmy Carter and the Iranian Hostage Crisis," http://www.whitehousehistory.org/teacher-resources/jimmy-carter-and-the-iranian-hostage-crisis

Chicago Tribune Editorial Board, "Benghazi: What the Report reveals About Hillary Clinton," http://www.chicagotribune.com/news/opinion/editorials/ct-benghazi-hillary-clinton-obama-rhodes-edit-0629-jm-20160628-story.html

Julian Hattem, "RNC Rolls out New Attack on Clinton emails," http://thehill.com/policy/national-security/272555-rnc-rolls-out-new-attack-on-clinton-emails

Chapter 6
The bin Laden and Guantanamo Take Down

Liz Sidoti, "McCain Challenges Obama on Terrorism," http://usatoday30.usatoday.com/news/politics/2008-06-18-1221196227_x.htm

Tom Basile, "The War on terrorism: Choosing to Fight the George W. Bush Way," http://www.forbes.com/sites/thomasbasile/2015/11/20/its-time-to-revisit-george-w-bushs-global-anti-terror-strategy/#1f9d5504782e

Unapologetic Liberal, "How to fight the Lie that Obama Doesn't Deserve Credit for bin Laden Death," http://addictinginfo.org/2011/05/04/how-to-fight-the-lie-that-obama-doesnt-deserve-credit-for-bin-ladens-death/

Bruce Weinstein, "Why President Obama was Right not to Say 'Radical Islamic Terrorism'," http://fortune.com/2016/06/13/obama-trump-clinton-radical-islamic-terrorism/

Eric Levenson, "Obama Describes Boston Marathon Bombing as 'Two Brothers and a Crock Pot'," http://www.boston.com/news/national-news/2015/11/16/obama-describes-boston-marathon-bombings-as-two-brothers-and-a-crock-pot

Chapter 7
Cuba Libre

Dan Roberts, "Obama Lands in Cuba as the first US President to vist in half century," http://www.theguardian.

com/world/2016/mar/20/barack-obama-cuba-visit-us-politics-shift-public-opinion-diplomacy

Earl Ofari Hutchinson, "What's Really Behind Obama's Cuba Move," http://newamericamedia.org/2014/12/whats-really-behind-obamas-cuba-move.php

Adam Taylor, "The history of absurd American plots in Cuba," http://www.washingtonpost.com/news/worldviews/wp/2014/12/11/the-history-of-absurd-american-plots-in-cuba/

"Presidential Candidates on Obama's Cuba Trip," http://ballotpedia.org/2016_presidential_candidates_on_President_Obama%27s_Cuba_trip

Charlie Savage and Julie Hirschfeld Davis, "Obama Sends Plan to Close Guantánamo to Congress," http://www.nytimes.com/2016/02/24/us/politics/obama-guantanamo-bay.html?_r=0

Chapter 8
A Presidential Conversion on Gay Marriage

Devin Dwyer, "Timeline of Obama's 'Evolving' on Same-Sex Marriage," http://abcnews.go.com/blogs/politics/2012/05/timeline-of-obamas-evolving-on-same-sex-marriage/

"Changing Attitudes on Gay Marriage," http://www.pewforum.org/2016/05/12/changing-attitudes-on-gay-marriage/

Ben Smith, "Obama backed same-sex marriage in 1996," http://www.politico.com/blogs/ben-smith/2009/01/obama-backed-same-sex-marriage-in-1996-015306

Chapter 9
Obama's Border War

David Savage, "Obama administration sue to black Arizona immigration law," http://articles.latimes.com/2010/jul/06/nation/la-na-arizona-lawsuit-20100707

Brian Tashman, "Five of the Craziest Reaction to Obama's Immigration Action," http://www.rightwingwatch.org/post/five-of-the-craziest-conservative-reactions-to-obamas-immigration-action-impeachment-race-war-rebellion/

"2010 State of the Union Address," https://en.wikipedia.org/wiki/2010_State_of_the_Union_Address

Tom Cohen and Bill Mears, "Supreme Court mostly rejects Arizona immigration law," http://www.cnn.com/2012/06/25/politics/scotus-arizona-law/

Julia Edwards, "Latinos furious at Obama on immigration delay, vow more pressure," http://www.reuters.com/article/us-usa-obama-immigration-frustration-idUSKBN0H20QX20140907

John L. Smith, "More heat than light among protesters at Del Sol," http://www.reviewjournal.com/columns-blogs/john-l-smith/more-heat-light-among-obama-protesters-del-sol

Mike Lillis, "113 Republicans Back lawsuit against Obama's immigration actions," http://thehill.com/home-news/house/241668-113-republicans-back-lawsuit-against-obamas-immigration-actions

Jim Acosta and Paul Steinhauser, "Bush welcomes new citizens in soft push for immigration reform," http://www.cnn.com/2013/07/10/politics/bush-immigration-speech/

Chapter 10
Trying to Outgun the NRA

Tim Devaney, "Obama and NRA on Collision Course," http://thehill.com/blogs/blog-briefing-room/news/264735-obama-nra-on-collision-course

"President Obama's 2015 Executive Orders on Gun Control," http://www.ncsl.org/research/civil-and-criminal-justice/summary-president-obama-gun-proposals.aspx

Daniel White, "Poll: Americans Support Obama Gun Control Initiatives," http://time.com/4173116/gun-control-barack-obama-polls/

"Gunned Down" The Power of the NRA," http://www.pbs.org/wgbh/frontline/film/gunned-down/

National Rifle Association, https://www.opensecrets.org/orgs/summary.php?id=d000000082

Dan Friedman, "Meet the Last NRA Democrat," https://www.thetrace.org/2016/05/last-nra-democrat-congress-2016/

Mike Spies and Ashley Balcerzak, "The NRA Placed Big Bets on the 2016 Election, and Won Almost All of Them," http://www.opensecrets.org/news/2016/11/the-nra-placed-big-bets-on-the-2016-election-and-won-almost-all-of-them/

Chapter 11
Thunder from the Left

Ralph Nader Interview, "Obama Trying to 'talk white' and appeal to white guilt," http://www.huffingtonpost.com/2008/06/25/nader-obama-trying-to-tal_n_109085.html

John R. McArthur, " Some Liberals Finally on to Obama's Betrayal of Liberalism," http://harpers.org/blog/2011/09/some-liberals-finally-on-to-obamas-betrayal-of-liberalism/

The Angry Independent, "Just Another War President," http://www.dailykos.com/story/2013/4/16/1202370/-Just-Another-War-President-Why-This-Progressive-Lost-Faith-In-Barack-Obama

Preston Maddock, "Faith Groups Oppose Obama's Drones Program," http://www.huffingtonpost.com/2013/04/16/faith-groups-obama-drones_n_3095443.html

Kevin Liptak, " How Obama came to launch strikes in Syria," http://www.cnn.com/2014/09/23/politics/obama-syria-strikes-policy/

Lucy Madison, "Ralph Nader: Obama is a 'con man'," http://www.cbsnews.com/news/ralph-nader-obama-is-a-con-man/

Chapter 12
Was a Black President Black Enough?

Michael Eric Dyson, "Barack Obama, The President of Black America?" http://www.nytimes.com/2016/06/26/opinion/sunday/barack-obama-the-president-of-black-america.html?_r=0

"The Obama Effect: African-Americans turned out last year in record numbers," http://datadrivendetroit.org/uncategorized/the-obama-effect-african-americans-turned-out-last-year-in-record-numbers/

Deroy Murdock, "Blacks Mainly Backslide under Obama," http://www.nationalreview.com/article/433222/black-americans-barack-obama-going-backwards-under-president-obama

Adam Hartung, "Want a Better Economy? History Says Vote Democrat," http://www.forbes.com/sites/adamhartung/2012/10/10/want-a-better-economy-history-says-vote-democrat/#18d1064667a1

Samantha Reyes and Kevin Liptak, "Obama issues 214 commutations for drug offenders," http://www.cnn.com/2016/08/03/politics/obama-drug-commutations/

"A Drug Policy for the 21st Century," https://www.whitehouse.gov/ondcp/drugpolicyreform

Carol E. Lee, 'Obama Says Police Killings of Two Black Men Should Trouble All Americans," http://www.wsj.com/articles/two-police-killings-of-black-men-part-of-pattern-of-racial-disparity-obama-says-1467919272

Julia Craven, "Here's How Many Black People Have Been Killed by Police This year (Updated)" http://www.huffingtonpost.com/entry/black-people-killed-by-police-america_us_577da633e4b0c590f7e7fb17

Devin Dwyer, "Obama on Race: We've Been Treated like the Help," http://abcnews.go.com/Politics/obamas-reflect-personal-incidents-racial-discrimination/story?id=27636612

Timothy M. Phelps, "Obama administration cracks down on police abuse, draws challenges," http://www.latimes.com/nation/la-na-obama-police-reform-20151022-story.html

Eric E. Sterling, "How Can Obama Fulfill His Pledges for Sweeping Criminal Justice Reform and Redeem His Policy Legacy?" http://www.huffingtonpost.com/eric-e-sterling/how-can-obama-fulfill-his_b_7899314.html

Chapter 13
In the Racial Cross Hairs

Benjy Sarlin, "Race surfaces in conservative protests against Obama," http://www.msnbc.com/hardball/race-re-surfaces-conservative-protests

Tom Jacobs, "Strong Link Between Race and Political Preference," http://www.alternet.org/election-2012/study-recent-elections-show-strong-link-between-racism-and-political-preference

'How Groups Voted in 2008," http://ropercenter.cornell.edu/polls/us-elections/how-groups-voted/how-groups-voted-2008/

Dominico Montanaro," Obama performance with white voters on par with other Democrats," http://firstread.nbcnews.com/_news/2012/11/19/15282553-obama-performance-with-white-voters-on-par-with-other-democrats

Kevin Liptak, "Obama: Because they marched, America changed," http://www.cnn.com/2013/08/28/politics/obama-king-speech/

Frank Newport, "Little 'Obama Effect' on Views about

Race Relations," http://www.gallup.com/poll/123944/little-obama-effect-views-race-relations.aspx

David Shepardson, "Obama says U.S. race relations have improved, but work to be done," http://www.reuters.com/article/us-obama-university-idUSKCN0XY0L8

Chapter 14
What to Do About the Poor

Lauren French, "Obama Swipes at Romney, GOP on Poverty," http://www.politico.com/story/2015/01/barack-obama-mitt-romney-poverty-114751

Peter Edelman, "Poverty in America: Why Can't We End It?" http://www.nytimes.com/2012/07/29/opinion/sunday/why-cant-we-end-poverty-in-america.html

Gillian B. White, "America's Poverty Problem Hasn't Changed," http://www.theatlantic.com/business/archive/2015/09/americas-poverty-problem/405700/

Leigh Ann Caldwell, "John Edwards says the Democrats should do more about poverty," http://www.cnn.com/2014/01/27/politics/john-edwards-sotu/

Sharon H. Chang, "The Growing Poverty Crisis that Everyone is Ignoring," http://thinkprogress.org/the-growing-poverty-crisis-that-everyone-is-ignoring-f903e9200d32#.g2m13gqpl

Bill Berkowitz, "Reject the History of Blaming the Poor for Poverty," http://www.truth-out.org/buzzflash/commentary/reject-the-history-of-blaming-the-poor-for-poverty

Danielle Kurtzleben, "Fact Check: Is It Obama's Fault

that Poverty Has Grown?" http://www.npr.org/sections/itsallpolitics/2015/08/17/432578282/fact-check-is-it-obamas-fault-that-poverty-has-grown

Chapter 15
Surviving the GOP's One Term Presidency Vow

Robert Farley, "President Barack Obama claims Mitch McConnell says his main goal is for GOP to regain the White House," http://www.politifact.com/truth-o-meter/statements/2010/oct/30/barack-obama/president-barack-obama-claims-mitch-mcconnell-says/

Andy Barr, "The GOP's no-compromise Pledge," http://www.politico.com/story/2010/10/the-gops-no-compromise-pledge-044311

Sophie Quinton, "Key GOP senators vow to oppose Susan Rice as Secretary of State," http://www.govexec.com/oversight/2012/11/key-gop-senators-vow-oppose-susan-rice-secretary-state/59516/

Paul Waldman, "GOP, Not Obama, to Blame for Divisiveness," http://www.realclearpolitics.com/2012/08/27/gop_not_obama_to_blame_for_divisiveness_288493.html

Jennifer Agiesta, "Bush now more popular than Obama," http://www.cnn.com/2015/06/03/politics/obama-approval-rating-cnn-poll/

Jennifer Agiesta, "Obama approval hits new high," http://www.cnn.com/2016/10/06/politics/obama-approval-rating-new-high/

The Obama Legacy

Timothy Noah and Cogan Schneier, "Trump poised to erase Obama policies," http://www.politico.com/story/2016/11/trump-erase-obama-policies-231156

"Republican and conservative support for Barack Obama in 2008," http://en.wikipedia.org/wiki/Republican_and_conservative_support_for_Barack_Obama_in_2008

Marc A. Thiessen, "The fall of the house of Obama is coming, and it's his own fault," http://www.msn.com/en-us/news/opinion/the-fall-of-the-house-of-obama-is-coming-and-it%E2%80%99s-his-own-fault/ar-AAkhJ1P?ocid=spartandhp

"Obama On Executive Actions: 'I've Got A Pen And I've Got A Phone'," http://washington.cbslocal.com/2014/01/14/obama-on-executive-actions-ive-got-a-pen-and-ive-got-a-phone/

Chelsea Bailey, "Obama says he'll give Trump a chance, even if he attacks legacy," http://www.aol.com/article/news/2016/11/21/obama-gives-trump-chance-even-if-he-attacks-legacy/21610827/?icid=maing-grid7%7Cmain5%7Cdl1%7Csec1_lnk2%26pLid%3D860364627_htmlws-main-bb

Bibliography

Abel, Daniel G. and Brown, Peter Harry, *Outgunned: Up Against the NRA* (New York: Free Press, 2012)

Alter, Jonathan, *The Center Holds: Obama and His Enemies* (New York: Simon & Schuster, 2014)

Austin, Algernon, *America Is Not Post-Racial: Xenophobia, Islamophobia, Racism, and the 44th President* (New York: Praeger, 2015)

Baim, Tracy, *Obama and the Gays: A Political Marriage* (Charleston, S.C: CreateSpace, 2010)

Baker, Peter, *Days of Fire: Bush and Cheney in the White House*, (New York: Doubleday, 2013)

Bowden, Mark, *The Finish: The Killing of Osama Bin Laden* (New York: Atlantic Monthly Press, 2012)

Bunch, Will, *The Backlash: Right-Wing Radicals, High-Def Hucksters, and Paranoid Politics in the Age of Obama* (New York: Harper, 2010)

Burton, Fred and Katz, Samuel M., *Under Fire: The Untold Story of the Attack in Benghazi* (New York: St. Martin's Press, 2013)

Dominguez Morales, Escoban, *United States-Cuban Relations: A Critical History* (New York: Lexington Books, 2008)

Draper, Robert, *Do Not Ask What Good We Do: Inside the U.S. House of Representatives* (New York: Free Press, 2012)

Dyson, Michael Eric, *The Black Presidency: Barack Obama and the Politics of Race in America* (New York: Houghton Mifflin, 2016)

Edelman, Peter, *So Rich, So Poor: Why It's So Hard to End Poverty in America* (New York: The New Press, 2012)

Emmanuel, Ezekial, *Reinventing American Health Care: How the Affordable Care Act will Improve our Terribly Complex, Blatantly Unjust, Outrageously Expensive, Grossly Inefficient, Error Prone System*, (New York: Public Affairs 1 Edition, 2014)

Graham, Otis L. Jr., *Immigration Reform and America's Unchosen Future* (New York: AuthorHouse, 2008)

Hafetz, Jonathan (ed.), *Obama's Guantánamo: Stories from an Enduring Prison* (New York: NYU Press, 2016)

Hashimi, Nader and Postel, Danny, *The Syria Dilemma* (Cambridge: MIT Press, 2013)

Hiltzik, Michael, *The New Deal: A Modern History* (New York: Penguin Books, 2011)

Hutchinson, Earl Ofari, *How the NRA Terrorizes Congress: The NRAs Subversion of the Gun Control Debate* (The Hutchinson Report Ebooks Kindle Edition, 2013)

Hutchinson, Earl Ofari, *How Obama Governed: The Year of Crisis and Challenge* (Los Angeles: Middle Passage Press, 2010)

Kinder, Donald R., *The End of Race?: Obama, 2008, and Racial Politics in America* (New Haven, Ct.: Yale University Press, 2012)

Leuchtenburg, *William E., In the Shadow of FDR: From Harry Truman to Barack Obama* (Ithaca, N.Y.: Cornell University Press, 2009)

Moats, David, *Civil Wars: A Battle for Gay Marriage* (New York: Harcourt, 2004)

Obama, Barack, *The Audacity of Hope: Thoughts on Reclaiming the American* Dream (New York: Vintage, 2008)

Press, Bill, *Buyer's Remorse: How Obama Let Progressives Down* (New York: Threshold Editions, 2016)

Press, Bill, *The Obama Hate Machine: The Lies, Distortions, and Personal Attacks on the President—-and Who Is Behind Them* (New York: St. Martin's Griffin, 2012)

Reid, T.R., *The Healing of America: A Global Quest for Better, Cheaper, and Fairer Health Care* (New York: Penguin Books, 2010)

Royce, Edward, *Poverty and Power: The Problem of Structural Inequality* (New York: Roman & Littlefield, 2008)

Schwab, William A., *Right to Dream: Immigration Reform and America's Future* (Fayetteville, Ark; University of Arkansas Press, 2013)

Smith, Jean Edward, *Bush* (New York: Doubleday, 2016)

Unger, Craig, *The Fall of the House of Bush: The Untold Story of How a Band of True Believers Seized the Executive Branch, Started the Iraq War, and Still Imperils America's Future* (New York: Scribner, 2007)

Woodward, Bob, *Obama's Wars* (New York: Simon & Schuster, 2011)

Index

J

K

M

N

O

About the Author

Earl Ofari Hutchinson is an author and political analyst. He is the author of best-selling books on politics and race in America. *The Obama Legacy* is the third in his trilogy on the Obama presidency. His other two works in the trilogy are: *How Obama Won* and *How Obama Governed: The Year of Crisis and Challenge.*

He is a contributing writer to the Huffington Post and is associate editor of New America Media. He is a weekly co-host of the *Al Sharpton Show* on Radio One. He is the host of the weekly *Hutchinson Report* on KPFK 90.7 FM Los Angeles and the Pacifica Network.

Made in the USA
Monee, IL
04 November 2026

67378667R00095